Published by the Canadian Electrical Association.

Design and Production: Presslink Communications Ltd., Montreal.

Typesetting: Alphabets Inc., Toronto.

Printing: Pierre Des Marais Inc., Montreal.

ISBN 2-9802153-0-9

Printed and bound in Canada.

Aussi disponible en français.

Legal Deposit – 4th trimester 1990.
National Library of Canada.
Bibliothèque nationale du Québec.

THE ELECTRIC CENTURY

An illustrated history of electricity in Canada

The Canadian Electrical Association
1891 - 1991

John Negru

CORPORATE UTILITY MEMBERSHIP

Alberta Power Limited
B.C. Hydro
Burlington Hydro-Electric Commission
Canadian Niagara Power Company Limited
Churchill Falls (Labrador) Corp. Ltd.
City of Calgary Electric System
East York Hydro
Edmonton Power
Etobicoke Hydro
Gloucester Hydro
Great Lakes Power Ltd.
Hydro Mississauga
Hydro-Québec
La Cie d'Énergie Maclaren-Québec
La Compagnie Hydroélectrique Manicouagan
Manitoba Hydro
Maritime Electric Company Limited
Markham Hydro Electric Commission
Montreal Engineering Co. Ltd.
New Brunswick Power
Newfoundland & Labrador Hydro
Newfoundland Power
North York Hydro
Northwest Territories Power Corporation
Nova Scotia Power Corporation
Ontario Hydro
Ottawa Hydro Electric Commission
Power Commission of the City of Saint John
Saskpower
Scarborough Public Utilities Commission
Toronto Hydro
Transalta Utilities Corporation
West Kootenay Power
Winnipeg Hydro
York Hydro
Yukon Electrical Co. Ltd.

PREFACE

It would be impossible to imagine Canadian society today without electricity. It is the energy behind many of the things we take for granted at work and at home and has been a driving force behind much of the growth of our country.

This book describes the history and growth of electricity and its use in Canada. It is not a dry description, or a chronology of technical milestones. Rather, it is a human and sometimes emotional look at how our society has progressed thanks, in part, to the magic of electricity.

As the Canadian Electrical Association and the utility industry enter our second century, we are taking this opportunity to look back at our achievements and accomplishments with pride. We are also looking to the future, undaunted by the complex challenges which await us; and confident in our ability to meet the needs and expectations of Canadians, with particular sensitivity for social and environmental considerations.

Wallace S. Read, President
Canadian Electrical Association

ACKNOWLEDGEMENTS

The support of all those involved in the production of *The Electric Century* has been invaluable. They include all CEA corporate utility and manufacturer members who supplied information for the research of this book and those who supplied photography, including Hydro-Québec, Ontario Hydro, Ontario Hydro New Business Ventures, TransAlta Utilities, The National Archives of Canada, Bell Canada Telephone Historical Collection, Notman Photographic Archives–McCord Museum of Canadian History, Canadian Marconi Company and Cinémathèque Québécoise, as well as the design group Presslink Communications and Alphabets Inc.

Those who deserve special thanks for their important contributions are Alan Wyatt, upon whose original research much of the book is based; the members of the CEA Centennial Committee, led by Chairmen Joe Drennan and Carl Kropp; Arlene Havrot, for her planning and the quality of her design work; Frank Bradley, who coordinated the project for CEA; and Alain Duquette for his sensitive translation.

CONTENTS

Transmitting towers of the Marconi Station in Glace Bay, Nova Scotia, 1902. *Courtesy of Canadian Marconi Company.*

INTRODUCTION

It was about 30,000 years ago that people first came to this place we now call Canada. More than 500 generations have passed since then, bringing with them many innovations and changes, but just about one hundred years ago, something happened that changed our lives perhaps more than any other event in our long history: the advent of electricity.

Looking around us today, it would be very difficult to find some place or aspect of our lives that is not touched by the force of electricity. It seems to be everywhere, from the plug in the wall to satellites in space. Yet we are the first generations in history to live entirely within the electric age.

Electricity. Where does it come from? How does it work? Many people never even bother to ask, even though it provides 25 percent of our energy. They just take it for granted: turn on the switch and there it is.

Even those who delve into its secrets find no simple answers to their questions. From amps to volts, and ergs to joules, the language of power retains a certain mystery to the average person.

Electricity is neither a finite natural resource, nor an unlimited right. As a manufactured energy form, it possesses unparalleled flexibility, but it can only be stored with difficulty. For the most part, it must be used up in the very moment it is produced, or the opportunity to use it is lost forever.

Our need for electrical power is so great, we have created huge power plants on far-off rivers in the North, and altered the very face of the land on such a scale that our efforts are visible from the moon. We have even bent the force of the atom to our will.

Yet no matter how little or how well one understands the nature and creation of this amazing thing called electricity, everyone will agree that it has transformed our lives and our world.

Out of the Past, Into the Future

It was in 1267 AD that Roger Bacon first published his theories on the polar attraction of magnetite. Almost six hundred years would pass before Michael Faraday demonstrated in 1821 that electricity could be generated by passing a wire through a magnetic field, and he went on to build the world's first dynamo in 1831. But once the power of electricity was harnessed, it swept the world like wildfire.

Within a few short years we had the telegraph (1846), the telephone (1876), incandescent lights (1879), streetcars (1883), the electric oven (1892), electric cars (1893), movies (1896), radio (1900), and a myriad of other inventions. Many of them came from Canada, and many more were to follow.

Thomas Edison, with an early light-bulb, circa 1880. *Courtesy of The National Archives of Canada*.

Until the mid-1800s, most of our power came from water, wood, wind, or brute force. Today, 99 percent of our work is done by machines.

It was only in 1853 that kerosene was invented (by a Canadian, Dr. Abraham Gesner, in Prince Edward Island), and lighting took its first big step from candles and whale oil into the modern age. But kerosene's heyday only lasted 28 years, until Thomas Edison perfected the incandescent lightbulb in 1881. Since then, lighting has been the exclusive domain of electricity.

Transportation was equally affected by the advent of electrical power. Many of Canada's first electrical companies were formed specifically to build and run tram-cars. By 1920, we had electric streetcars in 46 Canadian cities. But electricity's primacy in the field of transportation was fleeting. Now there are streetcars in only Toronto, although Light Rapid Transit systems in Calgary, Edmonton and Vancouver indicate a comeback for electrically-powered mass transportation.

The internal combustion engine did not have a big impact on transportation until well into the 1900s. In fact, oil first began to be exploited commercially in the 1860s. The first oil well in the world was drilled right here in Canada in 1858, in Enniskillen Township, Ontario, by James Miller Williams of Hamilton. Canada's first gusher went into operation, pumping 2,000 barrels a day, in Petrolia, Ontario, in 1862. For that matter, Canada's first real oil find (on the scale we think of today) didn't come until Alberta's Leduc No. 1 was brought in, in 1947!

We've gone from those first remarkable steps to a world of computers, car phones, instant replay, laser surgery, and satellite dishes in the backyard, all in the blink of an eye.

Now, we in Canada are the fourth-largest generators of electricity in the world – more than 482,158 gigawatt-hours in 1989. Our electric utility industry employs more than 90,000 Canadians and has assets of more than $100 billion. Several of our electric utilities rank amongst the largest corporations in the country, with service territories larger than many of the countries in western Europe put together!

Electricity in action today: monitoring the weather with computers. *Courtesy of Masterfile.*

We also boast some of the lowest electricity rates in the world, second only to Sweden, and less than half what our American neighbours pay. For a nickel, the average Canadian can light his or her home all evening. For an extra few cents, we can enjoy stereo music or a bit of TV too. A quarter will cook us a five-course dinner; and for fifty cents, we can invite some friends over and throw a party.

Manual labour played a major part in the establishment of utility transmission systems. This photo shows a Toronto Niagara Power Company work crew erecting a 12 kV line in 1907. *Courtesy of Ontario Hydro.*

Canada enjoys a unique position, where electricity is concerned. We have tremendous natural resources in rivers, coal, oil, natural gas, and uranium, from which to make power. As pioneers in electrical generation, transmission and distribution, we now possess a technological base, pool of knowledge, and efficient utility organizations that are world-class.

But Canada's harsh climate and the nature of our major industries have meant that we Canadians are also among the world's greatest users of energy. We have one of the highest ratios of energy consumed per unit of economic output among any industrialized nation on Earth. Japan, for example, uses one half the energy Ontario does, for an equivalent economic output.

Trying to describe our electric history, then, is more than just a simple matter of numbers, more than a recitation of how we got things done. It is a complex tapestry, with many skeins.

The Electric Century

The history of electricity in Canada is a truly fascinating story, a tale it is our great pleasure to tell in the centennial of the founding of the Canadian Electrical Association in 1891.

After looking through our past achievements, our thoughts naturally turn to the future, and we try to imagine what tomorrow holds. Electricity has changed our lives so much! But that is only half the picture. As Canada has come of age, we have undergone a dramatic shift in how we view our place in the world. We live in a global village, on a fragile planet.

For us today, understanding the role electricity will play in the lives of our children's children demands not only a grasp of technology and enterprise, but also a close look at the issues and dominant themes that shape our communal life — issues such as preservation of the environment, sustainable economic growth, partnership with the Third World, and spiritual values in a material world.

The Electric Century is more than an illustrated history of electricity in Canada. It is also a reflection on our aspirations and responsibilities, both as members of the Canadian electrical industry, and as individuals in the Canadian mosaic.

It is an invitation to you, to share with us the enjoyment of celebrating our heritage, to envision with us a brighter world, and to join in a remarkable journey.

The Canadarm helps us put satellites in space. *Courtesy of Superstock.*

"Any sufficiently advanced technology is indistinguishable from magic."

Arthur C. Clarke, from *The Lost Worlds of 2001*

CHAPTER ONE

IN THE BEGINNING

The first commercial application of electricity in Canada came with the formation in Toronto of the Toronto, Niagara and St. Catharines Electro-Magnetic Telegraph Company, in 1846. From that moment, inventions and applications poured forth with stunning speed.

When you consider that the dissemination of new information in those days was rather a hit-or-miss situation, it's quite remarkable to realize how quickly inventors and businesspeople latched on to the new possibilities of electricity. In fact, many of the early developments were specifically designed to enhance communications: by telegraph, undersea cable, telephone, and radio, thereby allowing the process to feed on itself and grow with ever-increasing acceleration.

For example, our first telegram went from Toronto to Hamilton on December 19, 1846 (just two years after telegraphy's introduction in the United States). By 1885, along with the completion of the CPR trans-Canada railway, we also had our first cross-country telegraph system. In December of that year, the first telegram was sent from sea to sea (New Westminster, B.C., to Canso, N.S.) in an astounding three minutes!

In 1852, we laid the first undersea cable in North America, from New Brunswick to Prince Edward Island, and by 1855, attempts were already underway to run a line across the Atlantic Ocean to England. That monumental task took eleven years and five tries before the first successful cable was landed at Heart's Content, Newfoundland, by what was then the largest ship in the world: the *Great Eastern*.

In 1874, a young man from Brantford, Ontario, shocked the world with his conception of a talking telegraph system. His name was Alexander Graham Bell. Two years later, on March 10, 1876, the telephone was born. Five months on, to the day, Bell first delivered the long-distance feeling – from Brantford to Paris, Ontario (with battery power from Toronto, using lines borrowed from the Dominion Telegraph Company).

Canada's first telephone exchange opened in Hamilton in 1878, with seven lines and fifty subscribers. Exchanges in Newfoundland, New Brunswick,

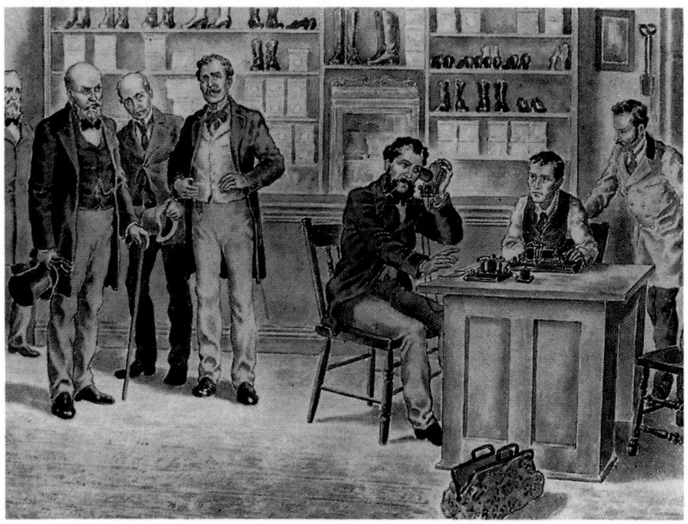

Scenes depicting the transmission and reception of the world's first one-way long distance telephone call between Brantford and Paris, Ontario on August 10, 1876. From paintings by J.C.H. Forster. *Courtesy of Bell Canada.*

Manitoba and British Columbia were also established that year.

Let There Be Light

Imagine, for a moment, that you are walking down a dark street on a moonless night, holding a candle. Perhaps the ground is knee-deep in snow, or the road is a labyrinth of muddy wagon tracks and icy puddles. It's tough going. Now imagine that someone has just turned on a street light with the power of 50,000 candles. Chances are, you like what you see.

That was exactly the reaction on March 12, 1873, when the Hon. Robert Davis, proprietor of the Davis Hotel on Main Street, in Winnipeg, switched on Canada's first arc lamp!

The event was reported by the Manitoba Free Press in the following words: "...the (electric) lamp in front of the Davis Hotel is quite an institution. It looks well and guides the weary traveller to a haven of rest, billiards and hot drinks, and lights up the streets more than the lamp of the newly incorporated gas company will for centuries to come."

And in spite of vociferous objections from gaslight interests, arc lights quickly found their way into cityscapes across the country.

Incandescent light, invented by Thomas Edison in 1879, also rapidly found its way into Canada. Rumor has it that one of his first demonstrations was held at McConkey's Ice Cream Parlour, in Toronto.

Within two years, mills on the Ottawa River were using lightbulbs, and in 1883, the Parliament Buildings were first lit up with power from a steam-driven reciprocating engine on the riverbank nearby. Our first incandescent streetlight system was inaugurated in Hamilton the same year.

When the lights in Saskatchewan went on in 1890, the Moose Jaw plant boasted its ability to power a whopping 350 lightbulbs (with a power of 16 candles each). The cost: $1.50 per bulb per month. By comparison, today's bulbs provide a combination of power and efficiency that makes them more than one hundred times cheaper per candle-power to operate!

Overhead wires on rue St-Jacques (formerly St. James Street), Montreal, 1896. *Courtesy of the Notman Photographic Archives, McCord Museum of Canadian History.*

It's for You...

Alexander Graham Bell had a daughter who suffered from profound hearing loss, and he had a lifelong mission of working with the deaf. The year after he invented the telephone in 1876, he returned to his first calling as a teacher of deaf children, and passed over 75 percent of his telephone patents to his father and Charles Williams, a Boston manufacturer who was to produce 1000 telephones. Williams was not equal to the task and quickly bowed out of the deal. Bell's disillusioned father decided to sell the company in 1879, but he couldn't find any Canadian takers. In 1880, he sold his interest to an American firm, National Bell.

It was James Cowherd, one of Alexander Bell's acquaintances from Brantford, Ontario, who started Canada's first telephone factory. Between 1879 and 1881, he made 2,398 phones.

Regina's first telephones were installed in 1883 – in the police station!

Canada's first dial phones were used on November 28, 1907, in Sydney Mines, Nova Scotia.

The first telephone, 1876. On the left is the receiver, on the right the transmitter. Early telephones could only send in one direction. *Courtesy of Bell Canada Telephone Historical Collection.*

A 1914 portrait o Alexander Graha Bell, inventor of the telephone. *Courtesy of Grosvenor, Bell Collection, Librar of Congress.*

All Aboard!

Electric street railways were another major innovation in Canada in the 1880s. Our first was demonstrated by the CEA's first president, John Wright, of the Toronto Electric Company, at the Toronto Industrial Exhibition (later known as the CNE) in 1883, a scant two years after their invention in Europe. The train consisted of three open flat cars with double rows of benches, pulled by an electric motor.

In 1886, Windsor got Canada's first public tramcar system. By the turn of the century, streetcars were a fixture in every major Canadian city, playing an important part in daily life. For example, when the tramcar operators in London, Ontario, went on strike in 1899, an ensuing riot by outraged riders required army intervention.

Electric tramcars on rue St-Jacques, Montreal, in 1907. *Courtesy of the Montreal Urban Community Society, Communications Department.*

Black(out) Comedy

Early understanding of electricity was sometimes hilariously absent. This inspired bit of reportage appeared in the *Welland Telegraph* during the summer of 1906, the day after a dynamo short-circuit had plunged the town into darkness:

"Something went wrong with the electric light service last night shortly after eight o'clock and for half an hour the darkness was turned on. The *Telegraph* called up the electric light station to enquire [into] the nature of the accident and was informed that the allamagoozalum had blown out. A reporter was soon on the spot and found Mr. Page and Mr. Houston wading knee-deep in the amperes of the electric fluid that had spilled out on the floor. The engineer lay stretched on the grass with several kilowatts of the allamagoozalum embedded in his neck. The allamag, etc., is a technical name which means load factor."

There were similar riots in Winnipeg in 1907.

Many of our early automobiles were also electric. The first appeared in Toronto on December 5, 1896. It was built for F.B. Fetherstonhaugh by Dickson's Coachworks, travelled at a maximum speed of no more than 30 km/h and lasted 15 years!

Electric cars were quite common, relatively speaking. But one has to remember that by 1903, when motor vehicle registration began, there were only 178 cars and trucks in Canada – all in Ontario. In 1905, New Brunswick became the second province to sport "horseless carriages" – twelve of them, in fact. Electric cars were around until the 1930s, when the internal combustion engine finally did them in. Now we have more than ten million vehicles on the road.

An early electric car. *Courtesy of the Bettman Archives.*

Get Cooking

The 1880s and '90s also saw the invention of the fractional horsepower motor and the introduction of many electric labour-saving devices: sewing machines, washing machines, stoves, toasters, coffee urns, milking machines, butter churns, and so on. Some, like the electric horse clipper, are no longer with us.

These innovations were not always well received. At least one Canadian preacher is on record as lambasting the new gizmos as: "...an invention of the devil [that] would release girls from honest toil to wander about the streets and fall prey to the wiles of Satan." Nowadays, the average Canadian home has as many as 80 electric appliances, from food processors to hedge clippers, and the list is still growing.

An electric stove was a big part of this "modern" kitchen, circa 1900. *Courtesy of the Bettman Archives.*

Power Up

Almost all of these miraculous inventions of the 1800s would not have been possible, had it not been for the development of corresponding facilities for the generation, transmission, and distribution of electricity.

The first independent electrical supply firm in Canada was the American Electric & Illuminating Company, established in Montreal in 1878. Since the gas companies had held a street lighting monopoly in Montreal and Quebec City since 1837, electric lights could only be found in private establishments, such as banks, hotels, and stores. It wasn't until 1886 that city politicians bowed to public pressure and agreed to install electric streetlights. Royal Electric Co. and Quebec Railway, Light & Power Co. won the tenders in the two cities, respectively, and in short order they each merged with their various electric and gas company competitors.

The first commercial utility in Canada was the Pembroke Electric Light Company, established in Ontario in 1884 and still operating to this day.

ABOVE: Power dam gates in Drummondville, Quebec, circa 1918. BELOW: One of the Montreal Light, Heat & Power Company's electric construction department trucks. *Both pictures courtesy of the Notman Photographic Archives, McCord Museum of Canadian History.*

Thanks, Brazil

"Abierte la Canada!" This Portuguese phrase from Brazil translates as: "Turn on the lights." This pun on the Portuguese word *candiero*, which means *electric candle* in English, was in common usage there around the turn of the century. The reason: São Paulo's municipal utilities were founded in 1899 by a Canadian company, the São Paulo Tramway, Light & Power Company (SPTLPC). The company went on to become one of Canada's greatest, Brascan. As for Brazil, they took over the SPTLPC in 1978.

Short and Sweet

"ALL RIGHT." That was the first message sent by Cyrus Field, over the new trans-atlantic undersea cable on July 26, 1866 (from Heart's Content, Newfoundland, to Valentia, Ireland).

"CANADA LINKED." That's the message HRH Queen Victoria received by cable on May 23, 1887, as Engine 374 steamed into Vancouver over the newly completed trans-Canada CPR line, with passengers from Montreal.

Courtesy of the National Archives of Canada, Gallaway Collection, Negative C-24989.

Running a utility in those days was not always an easy job. For one thing, profits were slim and competition was fierce. Companies came and went, were bought up, merged, or simply died, with dizzying speed. Equipment was limited, and blackouts were all too frequent. Utilities who shared the same poles would sabotage each other's lines with depressing regularity, and more than once police were called in to settle pugilistic confrontations between rival linesmen.

Yet through all this, the industry flourished. At first, power came from dynamos driven by steam from water heated by burning coal. Then, in 1881, the Ottawa Electric Light Company built a small waterwheel plant to supply power for 325 arc lamps for street lighting and 95 arc lamps for use in local mills. Their hydraulic generators at Chaudière Falls may well have been the first in Canada.

Water power, which had sometimes been referred to as "white coal," soon came to be known as "blue gold."

Hydro-electric power offered vast potential for meeting the growing public and industrial demand. The only problem was how to get the power from the river to the city. DC (direct current) electricity, which was the only kind invented at that time, could not travel very far without disappearing into thin air. It took the invention of the polyphase AC (alternating current) system by Nikolas Tesla, in 1888, to provide the solution. After a short period of controversy between the proponents

Looking north on Yonge Street, Toronto, in 1890. *Courtesy of the Notman Photographic Archives, McCord Museum of Canadian History.*

Ottawa Electric (Interior), Ottawa, Ontario, 1915. *Courtesy of the National Archives of Canada, Topley Collection, Negative E-130309.*

Nicholas Tesla (1856-1943) was an electrician and inventor who, in 1881, discovered the principle of the rotating magnetic field–the basis of practically all alternating-current machinery. *Courtesy of the Bettman Archives.*

of DC power, led by Thomas Edison, and those of AC power, led by Tesla and George Westinghouse, AC won the day. DC power didn't make a come back until the 1960s, when technological advances gave it new life.

The first three-phase AC plant in Canada went into operation in Calgary in 1889. The second was in St. Hyacinthe, Quebec, and came on line in 1895. It was located at Rapid Plat, ten kilometres out of town, and generated 450 kilowatts from three Canadian General Electric generators, at 2,500 volts.

The oldest high-head hydro-electric station in Canada was built in 1898, by the Cataract Power Company of Ontario, at DeCew Falls. It operated at 11,000 volts and the power was transmitted more than 56 kilometres to Hamilton.

AC had really arrived. And so had the 20th century.

Thomas Edison, inventor of the lightbulb, and champion of DC power. *Courtesy of the National Archives of Canada.*

"Canada, in fact, requires only to be known in order to be great."

J. Castell Hopkins, *Encyclopedia Canadiana*, 1898

CHAPTER TWO

POWERING THE NATIONAL DREAM

Much of Canada's history (and, indeed, much of our current cultural effort) has centred around attempts to link this country together – to forge one great nation. Many people think back to the trans-Canada railway of the 1880s in this regard. But that was only a physical and economic link. It took the combination of genius and electricity to really bring us together, and onto the world stage – with radio.

First, there was Guglielmo Marconi. He astounded the world in 1896 by sending a radio message more than eight kilometres, in England. In 1901, he came to Canada to perfect his invention. On December 12th of that year, he sent up a kite with an attached aerial, at Signal Hill, St. John's, Newfoundland, and received the first transatlantic wireless message (the letter "S" in Morse Code), from Poldhu, Cornwall, in England. About a year later, on December 5, 1902, he sent Canada's first wireless telegraph back the other way.

Marconi was a big hit in Canada, and his invention certainly seemed to epitomize the new spirit of the age. He also had substantial business acumen. Not only did he receive more than $23,000 in assistance from private and public funds (a fortune in those days); he also obtained an exclusive federal license to operate radio stations here.

But that is only one-third of the story. The second laurel belongs to Reginald Aubrey Fessenden, born in East Bolton, Quebec, in 1866.

Fessenden was also an inventor, and also interested in the idea of radio. A year before Marconi perfected his radio telegraph system, Fessenden did him one better: he sent the first spoken radio message, on December 23, 1900. Poor Fessenden (who, by the way, is credited with more than 500 other inventions as well) was shut out of the radio business by Marconi's monopoly, and never saw a penny for his radio apparatus until twenty-seven years later, when he got a $1 million settlement from the government of Canada.

The final luminary in the radio triumvirate was a whiz kid from Toronto called Ted Rogers. Someone in the United States had invented an AC vacuum tube

Marconi at Signal Hill, St. John's, Newfoundland, with the instruments on which he received the world's first transatlantic signal. *Courtesy of Canadian Marconi Company.*

(called the "Audion") in 1915, but it didn't work very well. Rogers is credited with inventing the first successful tube in 1925, and the world's first radio receiver that could be plugged into an electrical outlet.

Radio was slow to catch on at first. Our first commercial broadcast (consisting mostly of recorded music) took place in Montreal on November 4, 1919, over CFCF – the first radio station in the world.

By 1923, there were thirty-four radio stations operating around the country. Now there are more than one thousand, including the Canadian Broadcasting Corporation.

The Shape of Things to Come

Canada's population in 1900 was just over five million, of which the vast majority was rural. But our cities and our industrial capacity were growing quickly. The demand for electrical power continued to expand, and people were no longer satisfied with the existing system of independent, local, municipal, and private producers.

In Ontario, eyes began to turn toward Niagara Falls as a source of electricity on a large scale. Around the turn of the century, many of Ontario's electrical generating installations were powered with Pennsylvania coal. But in 1902, The Great Coal Famine (caused by strikes in the Pennsylvania coal fields), left the province with many factories closed, thousands out of work, and facing coal prices that doubled in the course of just one summer. There had to be a better way...

Four years of intense political debate followed, as government and business leaders struggled over who would develop Niagara Falls' hydro power, and build an electrical system for Ontario. Finally, in 1906, the government forces won the day, led by the charismatic Adam Beck, former mayor of London, Ontario: Ontario Hydro was formed – as the first government-owned utility in North America.

For the next eight years, the utility concentrated on transmission and distribution, buying power from other generating sources, and providing the public

Changing Horses

Early explanations of electricity's power tended to use images drawn from daily life. Light was measured in candle power, and motors were rated in horse power. But by the turn of the century, the new unified term "kilowatt-hour" was becoming the normative benchmark. Unfortunately, even today, most people have no clear idea of what a kilowatt-hour means. The conversion goes something like this: one horsepower is the equivalent of three-quarters of a kilowatt. Just multiply that by time and you have kilowatt-hours.

This list of domestic applications from a 1907 advertisement gives some insight into the power of a kilowatt-hour (as well as a peek into life in the "good old days"):

- Clean 5,000 knives, or 75 pairs of boots;
- Clip 5 horses;
- Warm your curling tongs every day in the year for 3 minutes, and twice on Sundays;
- Iron 30 silk hats;
- Light 3,000 cigars;
- Give you 3 Turkish light baths;
- Run a sewing machine for 21 hours;
- Carry you 3 miles in an electric brougham.

This 3-ton grain truck was part of the Adam Beck Circus in 1912 and featured electrical equipment ranging from a circular saw to a washing machine. *Courtesy of Ontario Hydro.*

with secure service at rock-bottom prices. In 1910, when the first Ontario Hydro transmission line opened, from Niagara Falls, to Berlin, Ontario (which was renamed Kitchener in 1914), the illuminated banner at the switching-on ceremony said it all: "For The People."

This Canadian pattern of government ownership of electric utilities was taken up by Nova Scotia and Manitoba in 1919, New Brunswick in 1920, then Saskatchewan in 1929, and Quebec in 1944. On the other hand, investor-owned electric utilities can be found in British Columbia, Alberta, Ontario, Prince Edward Island, and Newfoundland. Some work in tandem with government-owned utilities, others do not. Investor-owned utilities have proven themselves to be just as effective as government-owned ones, and share many of the same benefits.

On the technical side, projects got bigger, and also farther away from their end users. For example, in 1902, the Shawinigan Electric Company, in Quebec,

Sir Adam Beck, 1905. *Courtesy of The National Archives of Canada. Negative C-26070.*

Our Unsung Hero

"One, two, three, four. Is it snowing where you are, Mr. Thiessen? If so telegraph back and let me know." These were the first words spoken, on the first radio broadcast of a human voice, December 23, 1900. They were spoken by Reginald Fessenden, a Canadian inventor, at his laboratory on Cobb Island, Potomac River, near Washington, DC, over a primitive radio device and heard by his assistant Thiessen in Arlington, VA, 80 kilometres away.

Fessenden died in Bermuda in 1932. His epitaph there reads:

By His Genius
Distant Lands Converse
And Men Sail
Unafraid
Upon the Deep

Reginald A. Fessenden. *Courtesy of The National Archives of Canada, from the collection of the National Museum of Science & Technology. Negative PA-93160.*

installed the largest generators in the world (5,000 hp) and began sending the power some 135 kilometres to Montreal at 50 kV, along the longest transmission line in the world.

From 1900 to 1910, our overall installed hydro-electric capacity grew from 173,000 horsepower to almost one million horsepower. By 1920, it was up to two-and-a-half million. Over the next ten years, it sprang to more than six million horsepower!

A large part of that growth came from Ontario Hydro's first megaproject: the Queenston-Chippawa plant at Niagara Falls (later known as Sir Adam Beck No. 1), completed in 1921. It was the largest hydro project in the world, and made Ontario Hydro the largest utility in the world. At the time, its detractors hailed the project as unnecessary and uneconomical.

Transmission underwent similar development. In 1900, 20 kilovolts was a standard. By 1903 it was up to 50 kV, and by 1910 it was up to 110 kV. The first Canadian transmission at 220 kV took place over a Gatineau Power Company line from Paugan Falls, in Quebec, to Toronto, in 1928.

Many of the features of the Canadian electricity business as it exists today also came into being during this period of rapid progress...

For example, it was Ontario Hydro which first introduced the "promotional rate structure" as an incentive to industrial and municipal clients, in 1910. Under this system, the more electricity used, the cheaper the cost.

The splitting of many electric utilities into generators, transmitters, and distributors also hails from this period, with provincial utilities buying power from independent producers (and later producing it themselves), transmitting it to population centres, and then selling it wholesale to municipal utilities for distribution. In many provinces, however, one utility does it all.

The National Electrical Code, which first appeared in 1897, was adopted by the Canadian Standards Association in 1927, and approved by British Columbia, Saskatchewan, Ontario, Quebec, and Nova Scotia. Public adherence to this

Now That's Entertainment!

Nowadays, going to the movies is probably one of the most popular forms of entertainment in Canada. But did you know that our first movie theatre opened its doors in 1906, in a converted Montreal dance hall? The proprietor was one Ernest Ouimet, the theatre was called the Ouimetoscope, and the show was four minutes long.

The Ouimetoscope was such a success that in 1907 its owner demolished the old building and built a new picture palace on the site. It's still there – the longest-operating movie theatre in North America. *Courtesy of La Collection Cinémathèque Québécoise.*

code is what ensures the safety of all electrical applications in homes, businesses and industries, as well as within the electric utilities themselves.

Another extremely important business development of this period was the response of provincial utilities to the groundswell of public demand for rural electrification. Since its inception, the electrical industry had concentrated its efforts in meeting the demands of urban customers, who represented only 20 to 30 percent of the population in many provinces. As generating capacity and transmission expertise grew, and the

Louisburg, N.S., circa 1915. *Courtesy of the Notman Photographic Archives, McCord Museum of Canadian History.*

possibility of rural electrification became a reality, utilities began to turn their attention to the fulfillment of their provincial mandates to provide low-cost power to *all* citizens. It was Ontario which passed the first provincial Rural Hydro-Electric Distribution Act, in 1921. The process of wiring virtually every nook and cranny of Canada took a great deal of concerted effort and had a profound impact on rural people's lives. Today, almost every person in southern Canada has access to secure, inexpensive electrical power.

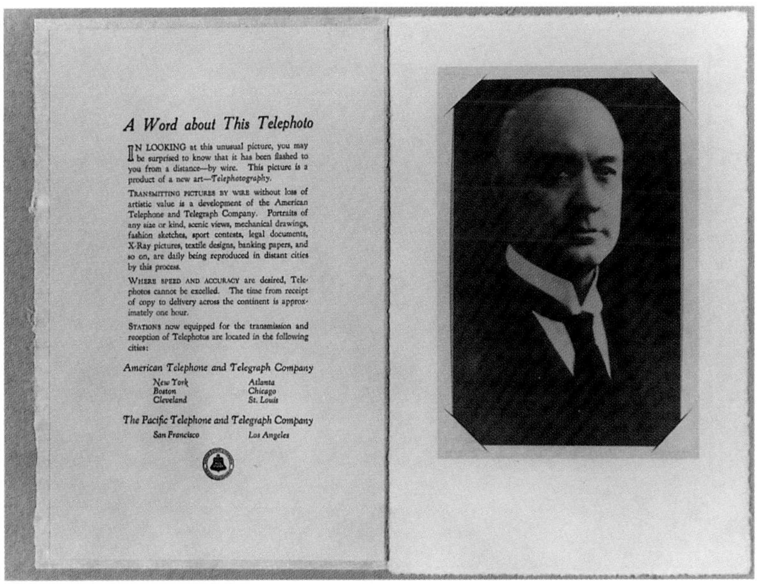

First, Biggest, Best

Nor were our achievements with electricity in the early part of this century limited to advances in generation and transmission. In 1906, Canada's first movie theatre opened in Montreal. In 1912, the Steel Company of Canada opened the world's first all-electric steel mill. In 1913 we installed the first electric reversing mine hoist here. We launched the world's first electrically-welded ship in 1918. And in 1919, Professor Ernest Rutherford, formerly of McGill University in Montreal, ushered in the nuclear age, with the world's first induced atomic reaction.

In 1924, the world's first newspaper photo transmitted by radio (using the invention of a Canadian, Sir William Stephenson) appeared in the *Daily Mail*, of London, England. In 1925, as mentioned earlier, Ted Rogers invented the batteryless radio receiver. In 1926, the Yukon and Northwest Territories opened their first successful communication link

An early advertisement for telephoto technology, circa 1925. *Courtesy of the Notman Photographic Archives, McCord Museum of Canadian History.*

with the south, by radio. (Earlier attempts to set up a telegraph link had succumbed to the rough terrain and climate.) In 1928, Morse Robb of Toronto invented the electric organ. In 1929, the first diesel electric locomotive in North America completed its maiden trip from Montreal to Toronto, and ended the hegemony of the steam engine.

In 1932, the trans-Canada telephone system was inaugurated, and in the same year the federal government passed a bill establishing a nationally-owned public broadcasting system. 1936 saw the world's first television broadcast (*Picture Page,* on the BBC) starring Joan Miller of Nelson, B.C. In 1939, Dr. James Hillier of Brampton, Ontario, and Albert Prebus of Edmonton, invented the electron microscope.

That's quite a record of achievement! As you can see, the world we know today is not so old...

Watching President Truman address the United Nations on TV in 1946. *Courtesy of the Bettman Archives.*

“The measure of a civilization is how much you take for granted.”

Louis Dudek, Canadian poet, 1975

CHAPTER THREE

A NEW GENERATION

Talking about the role of electricity in Canada's development has revolved around its effect on society up to this point. But looking at the years since the end of World War II, we need a new perspective: namely, the demands of society on electricity.

Canada has seen a dramatic, almost inconceivable, growth in the last forty-odd years. Our population has more than doubled (from just over eleven million in 1941, to nearly twenty-six million in 1988). The weekly wage of an average Canadian has gone from the 1939 level of $23.44 to almost $500. The number of cars on the road has increased tenfold in the same period. Ninety-nine percent of Canadian homes now have a television, 98 percent have a telephone. And the list of statistics goes on; in every sphere, Canada has been transformed.

The changes mentioned here are just quantitative. But the entire quality of our lives has been transformed as well. The emerging possibilities and expectations of Canada in these years of metamorphosis have had a fundamental impact on our electric utilities.

The Power of Power

In the forty-year period, from 1920 to 1960, the demand for electricity in Canada grew by about 7 percent each year, forcing utilities to double their grids sixteen times over. The public's thirst for electric power seemed limitless. Utility planners began to look ahead and shake their collective heads: how could they keep up? New generating capacity needed many years to go from plans to production, and the costs involved were staggering. Canada's electric obsession was quickly becoming the major item on provincial agendas.

How we used electricity was changing too, as we moved from the industrial age into the information age.

Ontario, as Canada's population and business heartland, saw the most radical change. Its industrial sector dwindled in importance as the new society took shape, falling from 67 percent to 37 percent of the overall economic mix between 1950 and 1980. The overall result was a phenomenal rise in residential and commercial demand throughout the province.

CFTO's booth under the CNE arch promoting the "Live Better Electrically" theme, 1961. *Courtesy of the CNE Archives, Gilbert Milne Photo Collection. Negative 52399.*

Commercial demand for electricity (especially in the retail trade and service sectors) jumped 600 percent from 1957 to 1974. Industrial demand (including pulp and paper, mining and smelting, chemical refining, and auto manufacturing) rose only 200 percent in the same period.

The growth of residential demand was equally startling, and "Live Better Electrically," seemed to be the baby boomers' loudest battle cry.

In fact, it was the arrival of western natural gas in eastern Canada, in 1955, that sparked the "Live Better Electrically" era. Gas cost 40 percent less than electricity, and the price was falling. Municipal utilities were worried: electric ranges and electric water heaters were among their most lucrative sources of income. In Ontario, they demanded that Ontario Hydro fight back against the gas interests.

Missionaries for the cause fanned out to women's clubs, schools, conventions, etc. across the province, with presentations such as Electricity's Story, Hydro's Homemakers, Sisters of the Skillet, Operation Eskimo (for home

Selling the future, early 1950's. *Courtesy of the Bettman Archives.*

Looking south on Yonge Street at Dundas, Toronto, 1951. *Courtesy of York University Archives, Toronto Telegram Collection.*

freezers) and Operation Dryer Drive (for clothes dryers). The public went crazy.

When Ontario Hydro unveiled the "Gold Medallion" all-electric home at the Canadian National Exhibition in 1958, more than 112,000 people jammed in to see it. At one Gold Medallion subdivision in 1963, 12,000 people visited the model home in the first weekend, and crowds of 3,000 per weekend were the norm for weeks afterwards.

Nor was this residential focus limited to the province of Ontario.

In Saskatchewan, *Penny Power* promoted electricity to homemakers through newspaper columns, personal appearances, and cookbooks.

Calgary Power was the first in North America to lay a prototype underground distribution system for residential subdivisions, in 1950, thanks to technical breakthroughs in using polyethylene insulation rather than rubber. The instant success of this pole-free innovation can be seen in virtually every postwar suburb in North America.

In other words, the ethic of limitless economic growth and conspicuous consumption that gripped North America in the 1950s and '60s was just

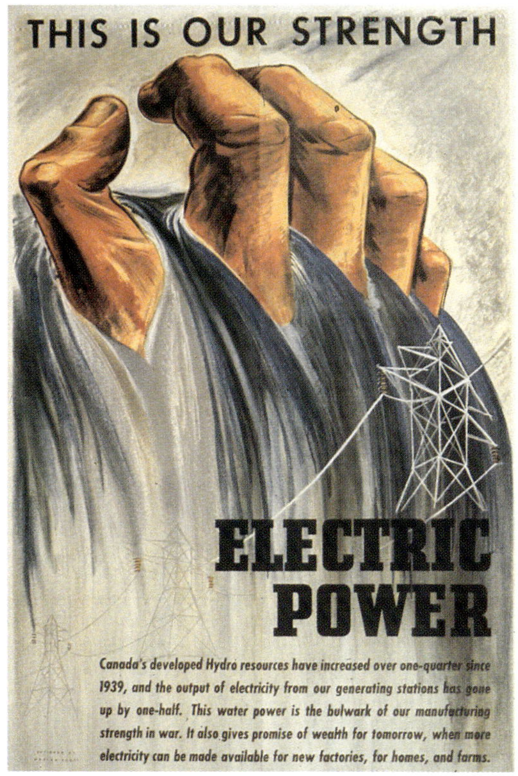

A World War II poster promoting electric power, used between 1939 and 1945. *Courtesy of the Metro Toronto Reference Library.*

Sky Power

SaskPower has been a leader in using the new technologies of satellites and fibre optics to run their generating plants with efficiency. In 1985, SaskPower began *Operation Skyswitch*, a remote control and communications program for its various generating facilities, using the Anik C-1 satellite and six earth stations around the province. In the same year, it also installed a fibre optic system for voice/control/protection signals from the Codette Switching Station to the Nipawin Hydro-electric Station.

Sea Power

On August 24, 1984, Nova Scotia Tidal Power Corporation commissioned the first tidal generating station in North America, at Annapolis. Its Straflo Turbine operates at 20 MW. The Bay of Fundy is another possible site for a tidal generating station, but design difficulties like silting and flooding have yet to be overcome before we will see wide application of this fascinating possibility.

as strong in the electrical field as any other. Some say electricity was even the prime ingredient. But where would the power come from?

Think Big!

It was the age of the megaproject.

The St. Lawrence Seaway project was one of the first. This mammoth international undertaking (involving, on the power-generating side, Ontario Hydro, Hydro-Québec and the New York Power Authority) took 30 years to negotiate and plan. It took four years to build, and was completed in 1958. In the process, we relocated 6,500 people from eight communities along the river, and built entirely new town infrastructures.

In 1963, Manitoba Hydro began work on harnessing the Nelson River, Canada's fourth longest. The main problem there, it seemed, was not how to generate the power, but how to get it 1,000 kilometres to the populated part of the province. The solution involved design from Canada, and equipment from the United Kingdom and Sweden. When Manitoba's first transmission line from

The spillway at La Grande II. *Courtesy of Hydro-Québec.*

Wind Power

In 1986, at Kuujjuaq, on the banks of the Koksoak River in northern Quebec, Hydro-Québec (in collaboration with Energy, Mines and Resources Canada and the communities of Kuujjuaq, Kativik and Nayumivik) completed a demonstration wind turbine electrical generating station.

Evaluations over the next two years showed 77,000 kWh of electricity produced each year, even though the turbine operated at 80 percent efficiency. Granted, this represented less than one percent of total annual local production, but the wind turbine showed significant promise. It operates at temperatures as cold as -47° Celsius, with a top-rated power potential of 65 kW, and with 98 percent reliability.

What does it look like? It's a vertical-axis Darrieus windmill. The rotor is 15 meters in diameter, there are three polyester/fibreglass blades and it weighs 20 tonnes.

As a demonstration project, Kuujjuaq cost $600,000. New installations would cost about half as much. Hydro-Québec is currently planning to build a wind park with three turbines in a remote area on the banks of Ungava Bay, as well as developing a coupling system for combined wind-diesel operations.

Power from the wind. *Courtesy of Hydro-Québec.*

the Nelson River powered up in 1974, it was HVDC (two lines of 450 kV) – the highest voltage and the longest line in the world.

Meanwhile, Hydro-Québec was developing the Manicouagan and Outardes Rivers. Transmission was a key factor there too, and the utility used the opportunity to test the world's first commercial application of a 735 kV AC line. This 1965 breakthrough in extra-high voltage (EHV) AC transmission involved technical contributions from more than ten countries. The line itself is a cluster of four cables, each a mere 3.5 centimetres thick, comprised of an aluminum shell around a steel core. But it is so efficient in reducing transmission power losses that it provides the equivalent of two 500 kV or four 330 kV lines!

In Labrador, Brinco (the British Newfoundland Corporation), of which Hydro-Québec owned a one-third share, established the Churchill Falls (Labrador) Corporation in 1965, to develop Churchill Falls. The success of Churchill Falls hinged on the sale of power to Hydro-Québec over three 735 kV lines. And although the Newfoundland government acquired Brinco in 1984, Hydro-Québec maintained its share in the Churchill Falls development.

Churchill Falls was to become the site of the largest single hydro-electric generating station in the western world, initially with 5,225 megawatts and later upgraded to 5,429 megawatts. In 1978 the La Grande II hydro installation, part of Hydro-Québec's James Bay development, surpassed the capacity of Churchill Falls before its upgrade, with 5,328 megawatts.

La Grande II is part of Phase One of the James Bay project, begun in 1972 and Canada's largest megaproject. All in all, it took 85,000 worker-years, 1,500 km of gruelling sub-arctic road construction, five airports, and $15 billion to build.

The James Bay Project covers more than 176,000 square kilometres. The area is home to more than 8,000 Cree indians, most of whom live along the shores of James Bay. Their villages include Chisasibi, Wemindji, Eastmain, Waskagheganish (Fort Rupert), Némiscau, Baie-de-Poste, Waswanipi and Whapmagoostoo (Poste-

THINK GLOBALLY, ACT LOCALLY

Lahore, Pakistan. *Both pictures courtesy of Ontario Hydro New Business Ventures.*

Canadian electrical know-how is valued around the globe. Utilities, consultants, and contractors lend a hand in about 8 percent of the world's international projects.

Not everyone is as fortunate as we are here in Canada, and it is important to remember that. People in three-quarters of the developing world do not have access to regular electric service. In rural Africa, less than 4 percent of homes have electricity!

Here is a brief rundown of Canadian electrical expertise at work around the world:

ASIA

India: Assisted with the Idukki Hydro-electric Project (780 MW) in Kerala, starting in 1960. Currently directing construction of the Chamera Hydro-electric Project (540 MW) in Himachal Pradesh.

Pakistan: Overseeing rural electrification in Baluchistan. Preparing feasibility studies for a 120 MW generating station at Taunsa Dam on the Indus River. Provided technical assistance for Tarbela Hydro-electric Project in 1981, and subsequent 500 kV transmission line to Lyallpur. Ongoing personnel training programs there and here.

Bangladesh: Providing a turnkey distribution system and training.

Thailand: Supplied five boilers (four at 150 MW, one at 300 MW) for coal-fired thermal stations. Currently providing three more 300 MW boilers for the Mae Moh station.

AFRICA

Egypt: Providing equipment and energy-management training to the Egyptian Electricity Authority for Cairo.

Kenya: Institutional strengthening and hydraulic engineering consulting.

Ghana: Developing a power system development plan since 1974 for the Volta River Authority. Now developing a National Electrification Planning Study for the Ministry of Fuel and Power.

Mali: Technical assistance with a 250 kilometre transmission line and distribution system for the Niger River basin. Also working on an international transmission network from Mali's Manatali Dam into Senegal and Mauritania.

Nigeria: Provided assistance with several thermal and hydro-electric projects. Providing voluntary assistance to St. Mary's Hospital for a feasibility study on a transmission line to the facility.

Cairo, Egypt.
Both pictures courtesy of Ontario Hydro New Business Ventures.

PACIFIC RIM

Korea: Provided engineering, construction, equipment, project management, and plant operation training for the Wolsung CANDU nuclear reactor.

Malaysia: Providing rate studies.

Singapore: Built a 2,000 MW thermal plant at Senoko.

Indonesia: Supplied four 400 MW boilers for the Suralaya coal-fired thermal plant, in addition to providing design and project management. Currently consulting on construction of a 500 MW underground hydro station.

New Zealand: Technical assistance with an undersea HVDC cable to link the north and south islands. Also consulting on reservoir stabilization in landslide-prone areas.

China: A five-company Canadian consortium is currently doing comprehensive feasibility studies for the Three Gorges Water Control Project (16,750 MW) in Sichuan. Another joint venture with Beijing is providing various types of generating and transmission equipment for the Chinese market. We are also supplying equipment for the Geheyan Hydro-electric Project, which will be the first power station to be built on the Quingjiang River.

Yangtze River, China. *Courtesy of Superstock.*

Belize City, Belize. *Courtesy of Ontario Hydro New Business Ventures.*

LATIN AMERICA

Colombia: Upgrading the distribution system in the city of Cali. Supplying equipment for the Alto Anchicaya Hydro-electric Project (340 MW).

Venezuela: Supplying equipment for the Guri Power Project (11,000 MW). Also helping with the related rural distribution grid.

Belize: Training for the Belize Electricity Board in transmission and distribution.

Nicaragua: Equipping a geothermal generating station.

Jamaica: Assisted with hurricane recovery activities.

Barbados: Consulting on engineering issues for diesel generators.

Haiti: Managing several utility upgrade and rural electrification projects.

ALTERNATIVE TECHNOLOGIES

Bigger has not always proved to be better, in the case of Third World development. Some huge hydro installations have quickly silted up due to planning oversights, and not met performance expectations. When the Aswan Dam was built in Egypt, for example, the consequent flow control stopped annual flooding down the Nile River. Farmers were hit "out of the blue" by $6 million annual fertilizer costs to replace what had previously been supplied by the Nile on a regular basis–free of charge.

The Mini-Hydel Station developed by an Ontario firm is a small-scale alternative that will alleviate this type of large-scale planning problem. It's cheap, simple to build and run. It's durable and has a huge potential market. China alone has more than 90,000 potential micro-hydro installations!

A photovoltaic company out of Ottawa has been exporting solar-cell manufacturing facilities to both India and China for some time. Electricity from solar energy is expensive, but in many locations it is the most practical choice.

In Newfoundland, a 300 kW wind turbine has been installed on Bell Island. This project was undertaken under the Enerdemo Canada Program which was enacted to demonstrate technologies related to renewable and nonconventional sources of energy.

One Toronto firm recently installed a 50 kW wind-turbine generator on the Turks & Caicos Islands, near Cuba. Canada is not itself a place particularly suited to wind power. We do have one large Darrieus (vertical-axis) wind turbine at Cap-Chat (4 MW), on Quebec's Gaspé Peninsula, and a small experimental wind farm in the North, but that's about it. But many other places do consider wind power as a good investment. California, for example, has lots of electricity generated on wind farms. However, wind power does hold a particular appeal for Canadians living in the North, where other methods of generating electricity are expensive and where noise pollution is not such a big factor as it would be in a crowded urban setting.

"If we are to have any hope of controlling the elements that will transform our lives, an understanding of science is imperative."

David Suzuki, Canadian environmentalist, 1978

CHAPTER FIVE

ON THE JOB

Early in 1990, Industry, Science and Technology Canada ran an advertising campaign in bus shelters around the country. The slogan read, "There's a future in Science and Technology. Un bel avenir en perspective. Les sciences et la technologie." One graphic shows two young people flying high through the air in a car with no wheels that appears to work like a jet. In the background are the outlines of gigantic buildings, and far below, a sports stadium is covered by a crystal-clear bubble.

Social commentators of the 21st century may look back on this as simply another example of "Future Past," visions of the future that were never meant to be. But whatever your vision of the future is, it most likely includes motors.

When we think of electric motors, our thoughts usually turn to the blender in the kitchen, or something like that. In reality, just under three-quarters of Canada's electricity consumption is industrial or commercial, not residential.

Motors are not the only thing electricity is used for in industry and business, but they do account for the biggest part: 73 percent. Process heat and electrolysis require 16 percent. Lighting comes in at 11 percent.

Industry

The demand for industrial process heat is on the rise in Canada. In some industries, like pulp and paper, it may account for as much as 25 percent of the energy bill. Currently, almost all of the heat supplied (95 percent) comes from burning fossil fuels. But research into electric induction heating and co-generation are changing that.

On the business side of the utility picture, electricity for industry is also the major component in revenues.

B.C. Hydro, for example, has 1.23 million customers. While only 85 of them are major industrial accounts, those clients account for 36 percent of the domestic provincial load. The sectors represented range through mines, pulp and paper, forest products, and manufacturing. Pulp and paper represents the largest segment, at 44 percent of industrial sales, and the smart money says it will reach 50 percent by 1999.

Getting Their Own House in Order

B.C. Hydro's "Resource Smart" program, initiated in 1988, aims to make existing facilities more efficient and thus obtain significant additional energy and capacity. As of March 1990 the program has made 380 GWh of energy available and by 1998, this program is expected to make 1,260 GWh available. This additional energy may be obtained from existing plants by such things as additional generating equipment, raising operating levels of reservoirs, diversion of additional water, and improvements to maintenance procedures.

Many Canadian companies and municipalities have each saved thousands of dollars annually by eliminating unnecessary lighting and replacing older lighting technologies with sodium lamps.

Small Is Beautiful

In 1988, Alberta passed the Small Power Research and Development Act, to promote the purchase of electricity from smaller producers, and to set fair sale prices.

In Nova Scotia, five percent of provincial generation is privately owned and operated, and an independent power producer interconnection policy has been in place since 1982.

Each province has its own unique combination of industrial, commercial and residential markets. But one thing is consistent across the country: Canada's industrial sector cannot exist without electricity, and lots of it! There is a direct relation between our enjoyment of the benefits of our industrial society and our commitment to making it work.

Commerce

In the Canadian office of the 1990s, energy efficiency is the motto of the day, as smart machines do more with less.

Commercial electricity use breaks down like this:

- 38.4 percent office machines & motors
- 37.4 percent lighting
- 13.1 percent cooling
- 11.0 percent heating (space & water).

When we went through our skyscraper period in the 1960s and '70s, we had a penchant for glass towers. They proved to be astoundingly inefficient in terms of energy management.

New architectural designs, such as the Ontario Hydro headquarters in Toronto, have given us super-efficient buildings that look nice and save a bundle. Some don't even need furnaces!

And businesses have found other ways to save money too. Toronto's Sheraton Hotel installed a computer network a couple of years ago to control light and heat in its rooms based on occupancy. The system cost $200,000 but saved the hotel more than $100,000 a year – providing an impressively short payback period.

Computerized energy management earned the Westmetro III tower CEA's 1989 Commercial and Industrial Building Design Award. *Courtesy of Inducon Development Corporation.*

“Got Any Bright Ideas?”

That’s the slogan of a poster campaign run by Energy, Mines and Resources Canada not too long ago. The purpose was to elicit suggestions on energy efficiency from Canada’s producers, and the general public. Care to contribute?

Courtesy of Energy, Mines and Resources Canada.

Smart Minds, Sharp Pencils

Many of the major developments of this century have involved electricity. We've highlighted some of them, especially those made in Canada. But so far, we've stuck more or less to the period before 1950. Now let's look at more recent inventions. While relatively new on the scene, their impact on our lives is already compelling.

Consider, if you will, the fax machine. Its market penetration went from zero to total in something like two years! You may think that the fax is frivolous. How about these innovations...

Computers

Whole books have been written about the social impact of computers, in any one of myriad ways: from banking machines, desktop publishing, and telephone networks, to robots.

This is the Information Age. We've got the global village, data management, and process control. Now the form of electric information is limited only by the horizon of function. Computers are here to stay. In fact, there are more than 100 million in use around the world today.

Banking is one service sector that has been revolutionized by computers, particularly with the introduction of automated banking machines (ABMs). There are more than 8,900 ABMs in Canada now. Many offer users the possibility of access to on-line banking services around the world, 24 hours a day! Nor are deposits and withdrawals the

From the shop floor to the executive suite, you would be hard-pressed to find a modern manager without a computer nearby. *Courtesy of Masterfile.*

An automated banking machine – one step closer to a cashless society. *Courtesy of The Royal Bank of Canada.*

only thing you can do on an ABM. In the 1990s you may well be able to buy stocks and bonds, make reservations for airlines, hotels and car rentals, or even buy tickets for entertainment and sports events, all on your local ABM terminal.

Then there are charge cards – almost 20 million of them in circulation here. They are accepted by more than 650,000 merchant outlets in Canada, and annual sales volumes with plastic power exceed $30 billion per year.

The good old telephone has a new face too. In 1985 we went cellular. There are more than 445,000 users of cellular phones in Canada now, and by the year 2000, we could well have more than two million.

Medicine

Two of the most widely-used medical inventions in the world today were developed in Canada, and they both owe their success to the power of electricity.

One is the pacemaker, developed in 1951 by Dr. John Hopps and the National Research Council. The prototype model worked off a car battery.

The other is irradiation therapy for cancer patients. The key to the success of this treatment was a machine, known as the "Cobalt Bomb." It was Dr. Harold Johns, working in Saskatchewan, and a team at Victoria Hospital, in London, Ontario, who came up with the protective housing, along with the method of controlling dosage exposure and aiming the rays.

Lots of other valuable medical inventions involve electricity too.

CAT (Computerized Axial Tomography) scanners are one very sophisticated example. Non-intrusive diagnostic eye scanners that work with ultrasound are another example. The Hospital for Sick Children in Toronto now distributes pocket computers to diabetic children, so they can regulate their insulin doses by analyzing simple on-the-spot blood tests.

And next time you see one of those doctor shows on TV, have a look at the background in the trauma unit or intensive care – it's bound to be crawling with electronic equipment. Quite a contrast to 1896, when we started using X-Rays...

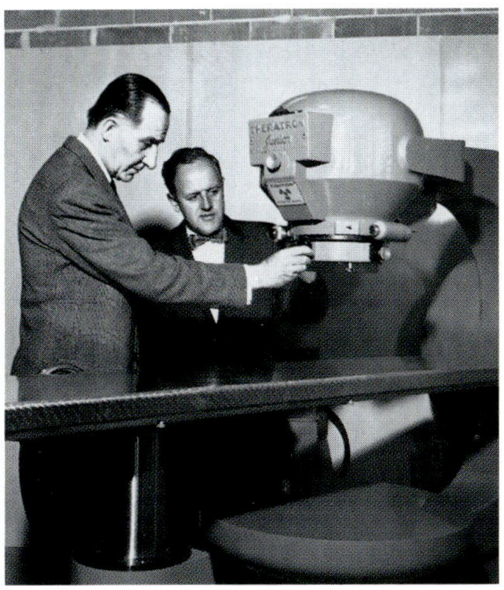

A Theratron Junior cancer therapy unit, Ottawa, 1959. *Courtesy of The National Archives of Canada, Photo Features/NFB Collection. Negative PA-163958.*

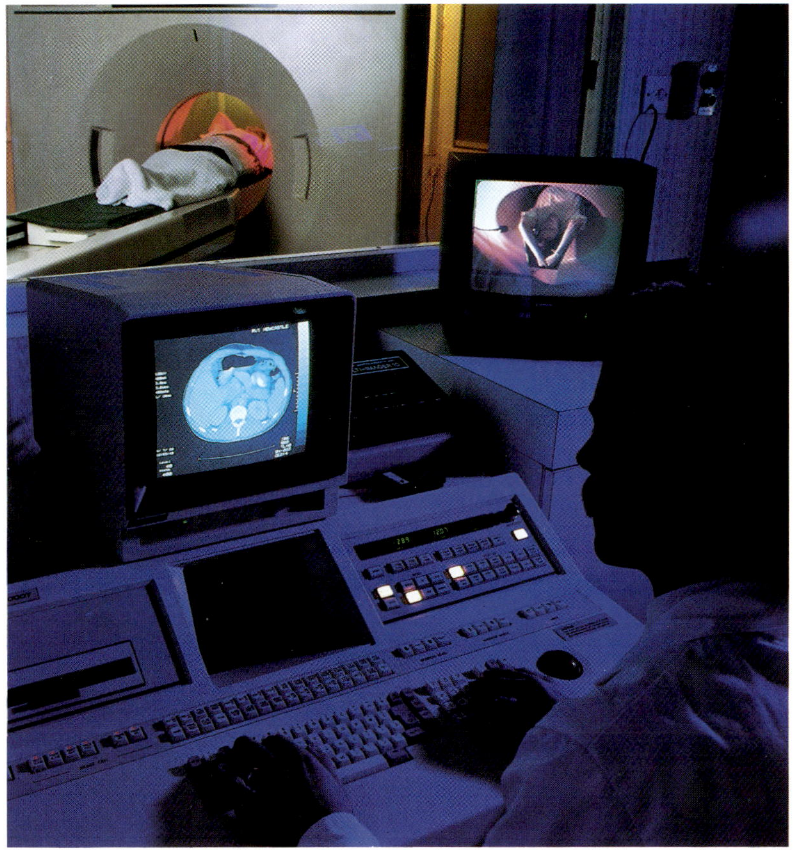

Computerized tomography scanning today. *Courtesy of Masterfile.*

Aviation

The President of the Royal Society in Britain, Lord Kelvin, is quoted as saying in 1885: "Heavier than air flying machines are impossible."

Co-generation for Cash Generation

Dow Chemical started a co-generation project in one of their Ontario plants, using residual process steam to drive several electricity-generating turbines once its first job was done. The move took them from being a huge electricity customer to being a net seller of power.

Increasingly, electric utilities are looking at co-generation as a means to meet future demand.

Courtesy of Grant Hill Photography.

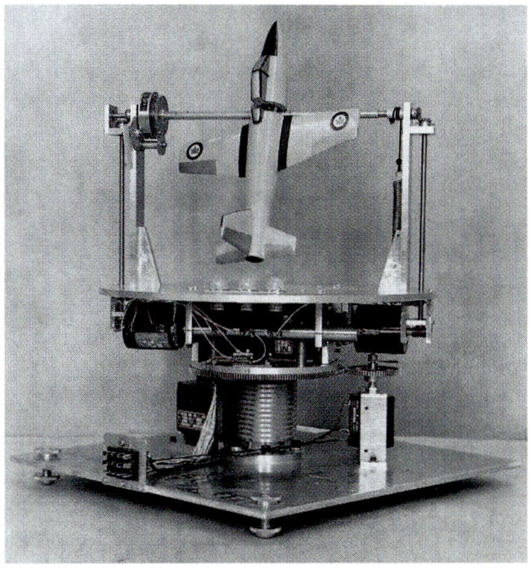

Canada's first flight took place on February 23, 1909, when the *Silver Dart* took off at Baddeck, in Nova Scotia. It was also a first in the British Empire. The first trans-Canada flight followed in 1920. It took ten days.

Now is the age of executive class, frequent flyer points, and last minute charters to the Caribbean and beyond.

It is true that we wouldn't have airplanes without the internal combustion engine, and our heavy use of airplanes would be different if there were no jets. But equally important is the electronic infrastructure that makes air travel so safe and easy. Three Canadian inventions that are integral to flight as we know it today were: the Position Home Indicator (1952); the Crash Position Indicator (1957), which went on to become the "Black Box," or flight recorder; and the R-Theta Navigational Computer (1958).

The case for the role of electricity in space is easily demonstrated by everything from satellites and the Canadarm to all those flight simulation games you can play on your home computer.

LEFT: A 1955 flight simulator. *Courtesy of The National Archives of Canada. Negative PA-167130.* RIGHT: A modern Boeing 767 simulator. *Courtesy of Masterfile.*

"*Nothing* is too visionary."

Sir Adam Beck, Chairman, Ontario Hydro, 1906

CHAPTER SIX

THE BUSINESS OF POWER

Canada has enjoyed many of the benefits of electricity thanks to the unique structure of our utility industry. We are faced with a huge land mass that is often inhospitable. Our population is concentrated in a thin band at the bottom, while our natural energy sources lie largely in the north. Within this framework, 83 percent of Canadian electricity is generated by provincial crown corporations or municipal utilities.

It's a very different picture in the United States. Only 21 percent of our neighbour's electricity is produced by government-owned utilities; 76 percent is produced by investor-owned utilities there.

This gives the United States one of the most diverse and complicated electrical industry structures in the world, including more than 1,800 municipal, 200 investor-owned, 70 public, 65 state-owned, and 35 federal electric utilities, along with more than 900 rural co-ops. Some of these companies are huge, but many are small and operate within narrow boundaries. This plethora of players, mandates, and hardware makes any large project, such as a major transmission grid, a protracted process.

In Canada, the structure of our electrical industry is much simpler, particularly because of our large provincial utilities. We have a mix of public and private companies, as well as smaller companies that operate on a municipal level, but the relatively small overall number of players makes things a lot easier. Some have argued that we should go one step further, and establish some kind of nation-wide power corporation, as most European countries have. But don't forget, any one of those countries could fit into one of our larger provinces several times over.

The Numbers Game

In 1989, Canadian utilities generated 482 billion kilowatt-hours of electricity. Sixty percent of it came from hydro power on average (primarily from Quebec, British Columbia, Newfoundland and Manitoba). Eighteen percent came from coal (with most in western Canada). Sixteen percent came from nuclear power, but all of it was in New Brunswick, Quebec and Ontario.

(Ontario relied on atomic energy for 46 percent of their electrical power in 1989.) Six percent came from thermal stations fuelled by oil or natural gas, or from soft-energy alternatives such as wind/solar/tidal technologies.

Hydro-electric power has provided a relatively smaller part of the mix recently because of a multi-year drought, resulting in greater use of fossil fuel–fired power plants. For example, during the winter of 1989-90, Hydro-Québec put its Tracy thermal generating station, near Sorel, into operation for the first time in a decade.

In 1988, Canada ranked fourth in the world in overall energy consumption, just behind the United States, the USSR and Japan. Per capita, we ranked second, just after Norway. We use 55 percent more electricity than our American neighbours, on a per capita basis.

Our electrical power is practically the lowest-priced in the world. Prices have tracked the Consumer Price Index fairly closely for the past 15 years. Oil, by comparison, has fluctuated wildly, although it is still the second most cost-effective choice after natural gas for home heating and similar applications.

In fact, the price of electricity in Canada (in constant dollar values) has fallen steadily over the years. In 1970 it was about one fiftieth of what it had been in the 19th century. More recently, prices have begun to rise again, but the increase is quite modest, compared to the overall benefits already achieved.

For example, when Hydro-Québec was formed in 1944, residential electricity prices in Quebec dropped 13 percent in just one month, and commercial rates went down 10 percent three years later. The rates didn't change until 1967.

Mix and Match: Monthly residential electricity costs – January 1988 (Dollars)*

43.48	Winnipeg	51.54	Edmonton	64.50	Moncton	101.70	Charlottetown
44.11	Montreal	53.32	Vancouver	70.58	St. John's	103.53	Yellowknife
50.12	Calgary	59.90	Toronto	72.30	Whitehorse		
51.20	Ottawa	60.73	Regina	72.61	Halifax		

* Bills computed are on a net basis, exclude all taxes, and are based on 1000 kWh. *Source: Statistics Canada, January 1988.*

Average 1988 world prices per kWh for residential electricity
(at 750 kWh annual consumption)

Prices in Canadian cents

Source: Energy, Mines and Resources Canada, 1988

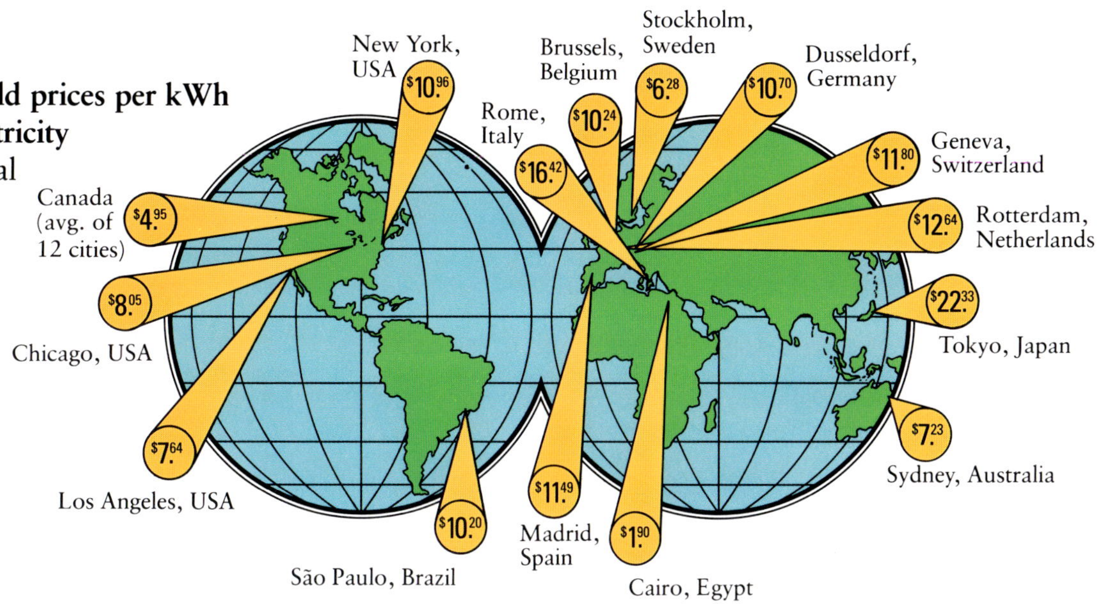

The Lay of the Land

In Saskatchewan 80 percent of electricity comes from thermal stations (73 percent from coal). Hydro-electric power accounts for 20 percent.

Nova Scotia's mix is 67 percent local coal, 23 percent oil, and 10 percent hydro (including power from the Annapolis Tidal Power Plant, the first tidal generating station in North America).

Quebec, Manitoba, and British Columbia each produces about 90 percent of its electricity from hydro power. In Alberta, it accounts for less than 4 percent.

Alberta's TransAlta is the largest coal producer in Canada. In 1989, the utility's Highvale, Whitewood, and Montgomery mines provided 15.5 million tonnes of coal – more than 20 percent of Canada's total output.

Although Ontario has relatively few major hydro sources left to develop, the province generates roughly as much hydro-based power as Newfoundland and more than every other province except B.C. and Quebec. Ontario's source statistics for 1989 were 46 percent nuclear, 24 percent coal, and 27 percent hydro.

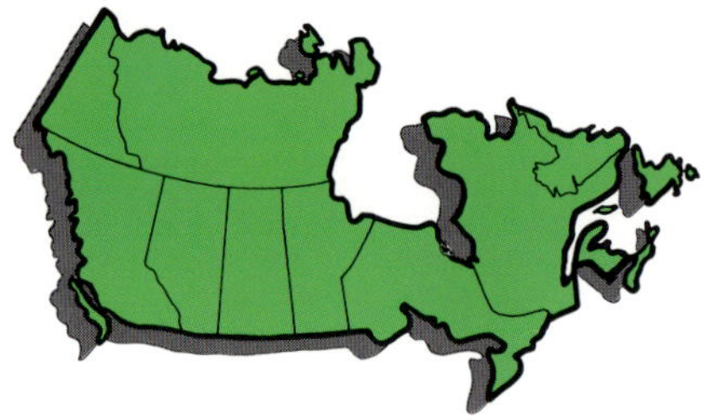

Going Public

Newfoundland

1954 Newfoundland Power Commission established
1965 Reorganized as Newfoundland & Labrador Power Commission (NPLC)
1974 Newfoundland government purchases Brinco's financial interests in Churchill Falls
1975 NPLC becomes Newfoundland & Labrador Hydro

Nova Scotia

1914 Nova Scotia Water Power Commission established
1919 Renamed Nova Scotia Power Commission (NSPC)
1937 Passes Rural Electrification Act
1969 NSPC completes purchase (begun 1945) of many private power companies including Canada Electric and Eastern Light & Power
1972 NSPC merges with Nova Scotia Light & Power Company to become Nova Scotia Power, a crown corporation, and takes control of all provincial electricity

New Brunswick

1920 New Brunswick Power Commission established

New Brunswick *cont'd*

1959 NBPC acquires distribution facilities of the Moncton Electricity & Gas Company
1969 NBPC completes purchase (begun 1947) of final portions of Maritime Electric's system in New Brunswick

Prince Edward Island

1977 PEI government acquires ownership of Northumberland Strait cable link to New Brunswick
1987 PEI government decides all future generating facilities and cables to be government financed and owned

Quebec

1910 Quebec Streams Commission established
1944 Creation of Quebec Hydro-Electric Commission
1963 Investor-owned utilities nationalized under Hydro-Québec

Ontario

1906 Hydro Electric Power Commission of Ontario (HEPCO) established
1912 Ontario Municipal Electric Association (OMEA) formed for commissioners

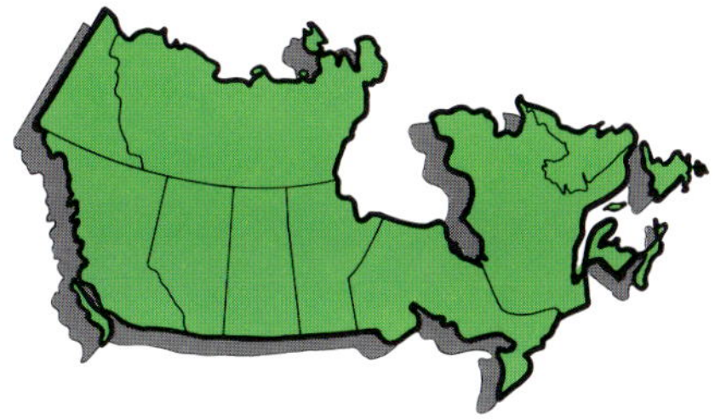

Ontario *cont'd*

1918 Association of Municipal Electric Utilities (AMEU) formed for managers
1921 Rural Hydro-Electric Distribution Act passed
1945 HEPCO completes purchase (begun 1922) of most remaining investor-owned utilities
1974 HEPCO renamed Ontario Hydro
1986 OMEA and AMEU amalgamated as the Municipal Electric Association (MEA) which now has 316 members

Manitoba

1906 City of Winnipeg Hydro-Electric System formed (and continues to this day as Winnipeg Hydro)
1919 Manitoba Power Commission established
1949 Manitoba Hydro-Electric Board (MHEB) formed
1952 MHEB acquires Winnipeg Electric, a major investor-owned utility
1961 Manitoba Hydro formed by amalgamation as a crown corporation

Saskatchewan

1929 Saskatchewan Power Commission established
1949 Becomes SaskPower, a crown corporation; Rural Electrification Act passed

Saskatchewan *cont'd*

1961 SaskPower reaches 58,000 farms and all villages/hamlets in the populated half of the province

Alberta

1982 Government creates Alberta Electric Energy Marketing Agency

British Columbia

1897 Investor-owned West Kootenay Power established
1945 British Columbia Power Commission (BCPC) formed
1962 BCPC amalgamates with B.C. Electric, the largest investor-owned utility in Canada at the time, to form the B.C. Hydro & Power Authority, known as B.C. Hydro

Northwest Territories & Yukon

1948 Northwest Territories Power Commission set up by federal government
1950 Legislation extended to include Yukon
1956 Renamed Northern Canada Power Commission
1988 Control passed to N.W.T. government, operating as the Northwest Territories Power Corporation

More Power...

In 1963, Hydro-Québec acquired most of the province's remaining private and cooperative electrical installations and began to integrate its diverse assets and operations into a new vision and a new, stronger company. With this control over all aspects of the production, transport, and distribution, Hydro-Québec saw the goal of bringing electricity to almost every corner of the province as finally attainable.

By 1988, Hydro-Québec was serving more than three million customers, of which almost 2,750,000 were in the domestic or agricultural sectors. The utility's industrial customers numbered a relatively small 13,368 (or less than one percent) but provided 29 percent of total sales volume in 1988.

The installed power of Quebec's generating stations at the beginning of 1989 totalled a tremendous 24,590 MW.

Fifty-three hydro stations provided 22,836 MW of this total (or 93 percent). Of these, the largest contributors were: La Grande II (5,328 MW), La Grande IV (2,651 MW), La Grande III (2,304 MW), Beauharnois (1,646 MW), Manic V (1,292 MW), Manic III (1,183 MW) and Manic II (1,015 MW).

Twenty-six thermal stations provided 1,754 MW of the total (or 7 percent). Of these, the largest contributors were: Gentilly II, nuclear (685 MW); Tracy, coal (600 MW); and La Citière/Cadillac, gas turbine (363 MW).

Shares of Canadian investor-owned utility companies are sought after by many managers of mutual and pension funds. *Courtesy of Superstock.*

Investor-Owned Utilities

Public utilities are not the only kind in Canada.

Newfoundland Light & Power (Newfoundland Power) was formed in 1966 when five electric utilities amalgamated to form the present company. In 1987 the shareholders of the company approved an arrangement to form a parent company known as Fortis Inc. whose shares are widely held across Canada by 8,000 common shareholders. The company supplies 85 percent of the population of Newfoundland.

Alberta relies on two investor-owned utilities: TransAlta Utilities and Alberta Power, who together supply 90 percent of the province's demand. TransAlta is the larger of the two, supplying 77 percent of the province's demand by itself.

TransAlta was formed in 1909, by Max Aitken, who was later known as Lord Beaverbrook. Today it is one of the most widely traded stocks on the Toronto Stock Exchange. Among their 23,500 investors, no single player holds more than 10 percent. As for Canadian content, a remarkable 99 percent of their shares are registered in Canada!

Alberta Power was formed in 1927. The shares of its parent company, Canadian Utilities, are held by more than 15,000 registered and non-registered shareholders, and represent a valued component in the stock portfolios of many Canadian mutual funds and pension plans, as well as individual Canadian investors.

West Kootenay Power in B.C. and Maritime Electric in P.E.I. are also investor-owned.

Power for Sale

The first international transmission line between Canada and the United States was built for power from Niagara Falls shortly after 1900. It operated at 12 kV.

The Strange Case of the Town of Summerside

Electricity came to Prince Edward Island in 1885. In 1917, the Maritime Electric Company was incorporated and operated in the province of New Brunswick. The following year the Company expanded its operations into Prince Edward Island by purchasing the Charlottetown Light & Power Company and moving its head office to Charlottetown.

During the early 1900s, several electric companies were engaged in the generation and distribution of electricity throughout the province. With relatively small and inefficient equipment and an ever-increasing demand for electricity, all of these small companies, with the exception of the Town of Summerside (population 8,000), found it uneconomical to continue independently. Repeated offers were made, and briskly refused by the town's citizenry.

The Town of Summerside continues to own and operate its own generating plant. Since 1961, the Town of Summerside has been interconnected into the Maritime Electric system, but it still keeps its diesel plant for standby and emergency generation.

In 1989, we exported 4 percent of our electricity south of the border (and passed another considerable chunk between the provinces). Sales to the U.S. brought $661 million into our economy.

By way of comparison, in 1988 we exported: 90 percent of our heavy oil; 78 percent of our uranium; 45 percent of our coal; 26 percent of our gas; and 20 percent of our light oil.

Ontario was the biggest electricity exporter until 1984. Now, Quebec, Ontario, and British Columbia are responsible for 70 percent of exports. New Brunswick picks up 24 percent and Manitoba about 6 percent.

The bull market has affected different regions in unique ways. For example, New Brunswick exported very little electricity until the late 1960s. Now it exports 25 percent of whatever it produces.

Before 1987, it looked to many people as though exports might be an increasingly important part of the Canadian electrical industry picture. But in the last few years, low water and increased domestic demand have combined to cut our export figures by a dramatic 60 percent.

Manitoba is the new big player in the export business, with hopes for major power sales from the Nelson-Churchill River hydro system. As for British Columbia: a connection with Puget Sound Power & Light is on tap for 1992, and with Washington Water Power for 1995.

Interprovincial transmission links play an important role in system reliability and efficiency. The future holds greater enhancement of reliability and efficiency through new opportunities for cooperative interprovincial planning. Nor should the scale of these achievements be underestimated, considering that practically all of our geography fosters North-South connections.

Transmission is the key link between generation and distribution of electricity. *Courtesy of TransAlta Utilities.*

“Land is not something you inherit from your parents. It is something you borrow from your children.”

Elmer McKay, MP, 1978

CHAPTER SEVEN

PROTECTING OUR HERITAGE

Canada's vision of itself has been through a lot of changes in the past hundred years. The fruition of many hopes and dreams, including electric power, has brought with it a new ethos... We know now that we cannot continue to enjoy Earth's natural resources limitlessly. Aside from the fact that these resources are finite, our very use of them has global consequences, in the shape of acid rain, the greenhouse effect, and hazardous wastes (from dioxins to spent nuclear fuel). Nor can we overlook the hidden social and environmental costs that follow upon any planning decision. Such calculations have become an integral part of all new energy ventures.

There is no one key for solving the many sensitive issues about which our new vision of the future has made us aware. We are caught, as it were, between the rock of demand for electricity and the hard place of harmful environmental impact.

Looking around at the rapid changes that the countries of the world have undergone just in the last one or two years, it's hard to imagine how the world will be in 25 years. Many developing countries see the benefits that come from increased electrical power as vastly more important than the possible environmental side effects. Our challenge is to continue to help these countries improve their quality of life without compromising our commitment to protecting the environment.

Although we in Canada have no way of shipping our electricity anywhere except the northern United States, our electrical industry and government have found other ways to assist in global development: through technology transfers, financial support, and training.

We are also making many vital breakthroughs in technology that will have environmental benefits, both at home and around the world.

Canadian utilities are focusing on several specific targets –

Demand-Side Management: Demand-side management involves actions intended to influence the amount or timing of a customer's use of electricity. The net impact is a reduction in demand that on a large scale can forestall the need for increases in electrical generating capacity.

The future of Canada is ours to enjoy, but also to create. *Courtesy of Masterfile.*

This includes load shifting, non-utility generation (including co-generation), and increased efficiency of equipment.

Emission Control: Burner-tip technologies such as FBC (Fluidized Bed Combustion), LNS (Low NOx SOx), and ICGCC (Integrated Coal Gasification Combined Cycle) are examples of improvements for coal-fired generating stations.

A lot of work has been going on in western Canada. Fortunately, western coal is much lower in sulphur than eastern coal. Alberta's electrical industry is committed to meeting the 1994 emission targets set by Canada and the province. The province's SOx emissions dropped by about 50 percent between 1970 and 1985. Thanks in part to TransAlta's pioneering LNS and IGCC technology, the future looks even brighter.

SaskPower will be the first North American utility to use LIFAC (Limestone Injection into the Furnace and Activation of the unreacted Calcium) technology developed in Finland to reduce sulphur dioxide emissions. SaskPower estimates that it could save $20 million over alternative emission control systems.

In New Brunswick and Nova Scotia, where existing coal has a higher sulphur content, all emission control technologies are being investigated.

Pollution Control: Management and safe destruction of existing hazardous chemicals (like dioxins or nuclear waste), finding replacements, planning for thermal pollution from processed water, reclaiming mined-out land, managing habitats, forestalling water shortages, investigating possibly-harmful biological effects of electro-magnetic fields... It's a long list.

A prototype LNS burner developed by TransAlta Resources. *Courtesy of TransAlta Utilities.*

Solutions Are Not Bought With Small Change

Between 1975 and 1985, the Canadian Electrical Association sponsored more than $10 million worth of research projects annually. In fact, research accounted for more than 20 percent of its total budget.

Between 1983 and 1986, CEA sponsored fourteen studies on acid rain, at a total cost of $3.6 million. Much of it concentrated on burner-tip technologies rather than more costly and complicated stack applications.

In 1986 alone, Canadian electrical utilities as a group invested $125 million in research. The CEA co-op research program (funded by utilities, government, industry and universities), spent an additional $15 million.

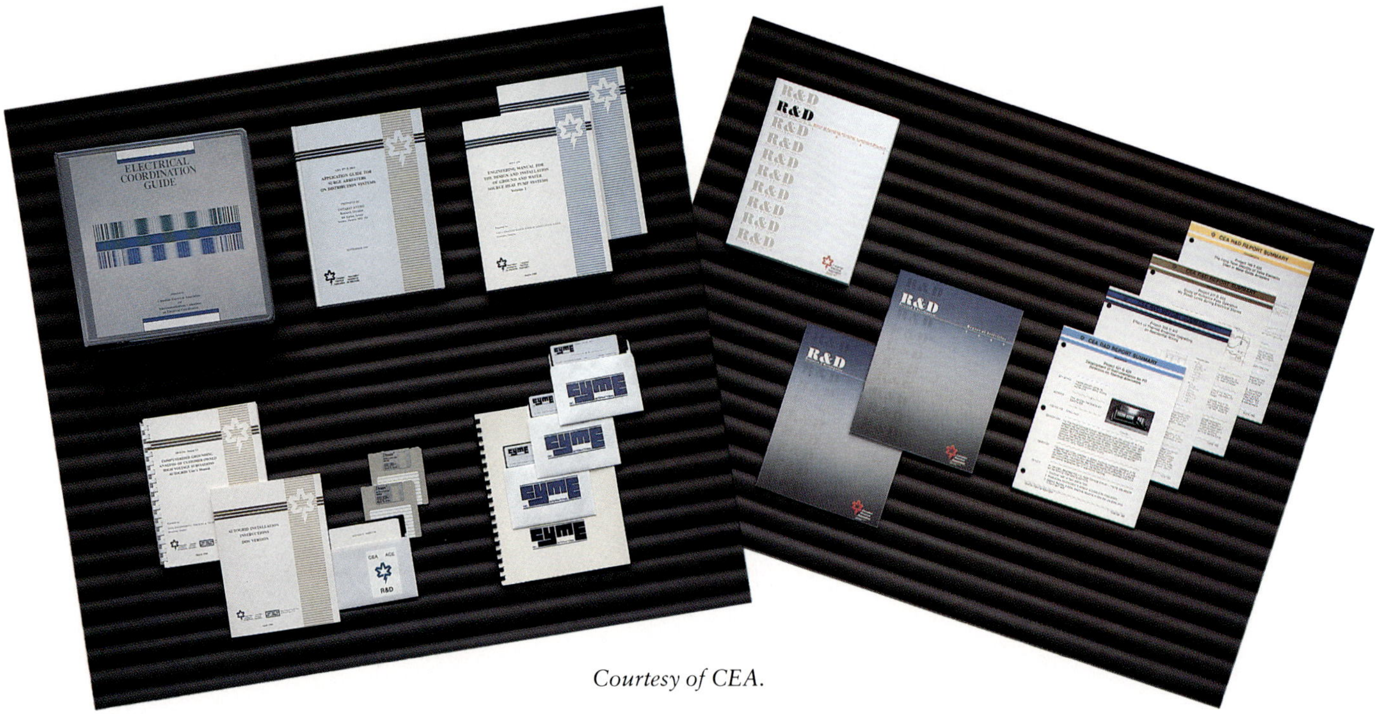

Courtesy of CEA.

Alternative Sources: These include redevelopment of existing generating facilities, additional sites on previously harnessed river systems (from large-scale projects like James Bay Phase II to micro-hydros), wind, solar, and tidal power.

Aside from hydro, wind power has the edge over solar, and tidal generation these days. The wind turbine generating station at Cap-Chat, on the Gaspé, rated at 4 MW, is our biggest example. The tidal station at Annapolis, in Nova Scotia, also generates about 4 MW. But for the most part, these soft-energy alternatives have yet to reach the stage where they can provide significant contributions.

In some parts of Africa, people spend eight hours a day getting water in jugs.

Gulp!

Canadian water consumption runs to about 380 litres per day per person:

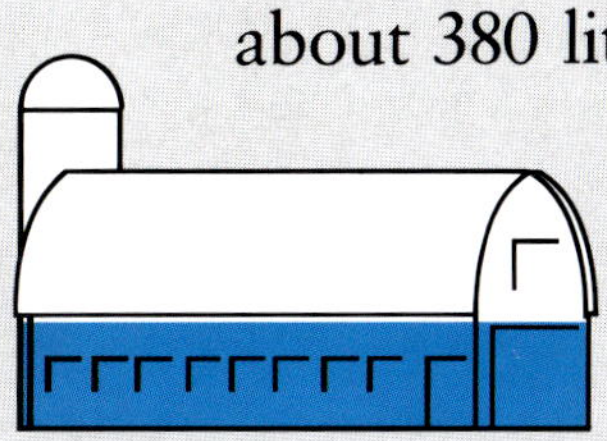

48 percent is used in agriculture;

22 percent is used in homes and offices;

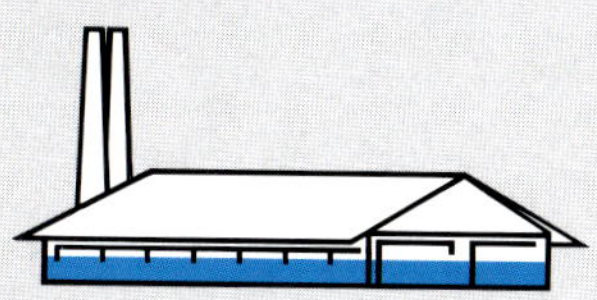

26 percent is used in industry;

4 percent is used in power station cooling.

The world is awash in water, but less than one percent of it is usable. The rest is too salty, too far, too cold, or otherwise inaccessible.

A typical 500 MW fossil-fired electrical generating station can circulate as much fresh water in a day as people's homes use in a city of 30,000. In other words, emission-gas control is not the only factor we have to consider: water quality is also important.

In many parts of the country, we use the same water over and over. Our hydrological cycle means we have dry spells every twelve years or so: water shortages are a very real concern in Alberta, Saskatchewan, Manitoba, Ontario, and Quebec.

Protecting water resources, then, is another key issue for utilities.

Unlocking the Atom

It was in 1923 that Robert Millikan, the Nobel Prize winner for Physics, remarked that: "There is no likelihood man can ever tap the power of the atom."

A mere 34 years later, Canada cut the ribbon on its first nuclear research station in Chalk River on November 3, 1957.

In 1962, Robert McCauley, the vice-chairman of Ontario Hydro, remarked that: "The nuclear horizon is unlimited... Recent developments make me confident that the day will come in our lifetime when electricity will be generated directly from an atomic unit the size of a water glass, without moving parts or the use of steam."

Canada's first CANDU nuclear generating station, Rolphton, on the Ottawa River, opened the same year.

No other technology has engendered more debate and strong feelings than nuclear power. Some see it as a dream come true, others as a nightmare waiting to happen. There are some 430 nuclear generating stations around the world, providing about one-sixth of our global electricity. Canada's CANDU technology is considered by many to be the best of the bunch. But aside from scientific impetus, why did we embark on a nuclear future?

Part of the answer lay with Canada's burgeoning demand for electricity in the

1960s. When planners looked at the growth trend in those years, it seemed the rise would never stop. At that rate, we would have had to double our generating capacity every ten years. By the year 2000, we would have needed sixteen times more power than was available in 1960. Hydro power could never have met that demand.

As if that weren't incentive enough, the OPEC oil embargo of 1973 sent oil prices on a roller-coaster ride, and strikes in the United States sent coal prices skyrocketing as well, ending any hopes for a secure, cost-effective electrical future powered by fossil fuels.

Ontario, Canada's largest consumer of electricity, saw itself as particularly vulnerable because it has relatively little undeveloped hydro potential and practically no coal. But one thing Ontario did possess was lots of uranium.

The control room at Darlington Nuclear Generating Station, Ontario's newest. *Courtesy of Ontario Hydro.*

Research Pays Off...

The Canadian Electrical Association's ongoing research and development program is a mammoth effort. Between 1985 and 1989 the CEA funded more than 325 projects, worth $60 million.

In 1989, for example, the strategic thrust of R&D covered environmental protection, improved safety, increased productivity, cost control, and understanding customers.

Projects in the environmental area in 1989 included studies on the greenhouse effect, small hydro and alternative energy, electric and magnetic fields, safe working clearances, wood pole preservatives, and overhead versus underground distribution systems.

Studies connected with customer service and demand-side management included: ground source heat pumps, electro-osmosis, microwave rendering, appliance efficiency, demand-side management modelling, and evaluations of customer acceptance of demand-side management.

On the generation and distribution side, research and development programs looked at trace chemicals in transformer oil, hydro generation operations planning, post-tensioned concrete, thermal plant life extension, metal oxide surge arresters, and hydro generation station upgrades.

Improving the delivery of electricity has been the focus of investigations into harmonics, system losses, fuse failures due to lightning, cost of interruptions, power quality standards, and predicting equipment life.

In 1990, the CEA had 179 ongoing projects with a total value of almost $39 million.

The benefits of this work for Canadian electrical utilities and the Canadian public, since the CEA R&D program was created in 1974, have made it worth every penny.

Thus it was that Ontario embarked on an ambitious plan to follow the lead of other countries, such as France, which have embraced nuclear technology as a viable solution for electrical power.

Now, there are 16 CANDU units in Ontario, with four more under construction. Nuclear power provides roughly 46 percent of the province's electricity.

Quebec and New Brunswick are the two other provinces where nuclear power is used. Hydro-Québec operates Gentilly II, and New Brunswick Power operates Point Lepreau.

The Chalk River Nuclear Research Centre, Canada's first.
Courtesy of Atomic Energy of Canada Limited.

"Everything that can be invented, has been invented."

Charles Duell, director, US Patent Office, 1899

CHAPTER EIGHT

INTO THE 21st CENTURY

Our future is not limited to what we can imagine, fortunately. It is the things we can't imagine that make life so exciting, for better or worse.

Still, we want to be the masters of our own fate, and that is as it should be.

Canada in the year 2020 will not be an electronic utopia. Some planners say we will run out of electricity by exceeding our installed capacity somewhere between 1993 and 2008. Plants wear out. At least 25 percent of Ontario's will be retired by 2014. We must think ahead now, because it takes up to fourteen years to approve, plan and build new sources of electricity.

Even with a 2.7 percent annual growth rate in demand for new electricity, we are looking at compounded increase in demand of almost 100 percent over the next 25 years.

Increased efficiency on the part of consumers and demand-side management by utilities are expected to meet 25 percent of the new demand. And what about the rest? Some provinces will go the route of increased imports. Others will redevelop or augment existing hydro-electric installations. Ontario plans to supply 53 percent of its new demand with nuclear power for base load and gas-fired thermal plants for peak periods.

Making the Future Bright

People, power, progress. Canada's electrical industries are constantly shining a bright light into the dark corners of the future. Research topics include generation, transmission and distribution, premium pricing for reliability and quality (in computer applications, for example), spin-off industries, demand-side management, load-shifting, and "soft" options.

Indeed, the quest for increased power efficiency has provided our economy with many golden opportunities. Canadian businesses, industries, utilities, governments, and universities are all responding to the challenge with inventiveness and vigour.

One technology that has some people sitting up in their chairs is hydrogen fuel cells. The idea goes something like this: We use electricity to break down water into hydrogen and oxygen. A few years ago Noranda Labs, in Quebec, developed a unipolar electrolysis system that is 85 to 90 percent efficient. We release the oxygen into the atmosphere, with our best wishes, and stockpile the hydrogen. Later the hydrogen could be burned to drive steam turbines and recreate electricity at a comparatively low cost and zero pollution at the generating site.

Hydrogen could also find its way into other sectors of the economy because, unlike electricity, it can be used as a fuel feedstock. Some see hydrogen-powered cars just like today's propane ones. But others point to the huge infrastructure that would be required to effect such a conversion. It is true, however, that the acceptance of battery-driven cars also hinges on a big infrastructure, not to mention the fact that battery technology is still not equal to the task. Electric railways stand a much better chance of widespread application than battery-based systems. Ontario's UTDC (Urban Transit Development Corporation) is a leader in this field.

One huge potential for hydrogen – and one which makes this technology so appealing to any country that has some serious hydro power – is that it is an essential ingredient in ammonia, which is an essential ingredient in... synthetic fertilizers.

Vancouver's LRT is one example of the new breed of public transportation systems that have made electrically-powered vehicles a viable option once again. *Courtesy of Superstock.*

Industrial Horizons

Electricity is also becoming a more and more important component of tomorrow's workplace, and new industrial technologies are predominantly electric. This wide-ranging list includes:

- heat pumps;
- variable speed motors;
- automated manufacturing;
- robotics;
- advanced control systems;
- induction and electric-arc furnaces;
- microwave and plasma heating.

Some of electricity's advantages in industrial applications are:

- compatibility with automation (CAD/CAM and robotics);
- providing faster, cheaper heating by microwave or induction, with lower thermal stress;
- cost-efficiency;
- price stability versus fossil fuels;
- no end-user pollution;
- smaller plant sizes;
- conservation of resources.

Electric-arc furnace. *Courtesy of Hydro-Québec.*

Precision welding with robotics. *Courtesy of ABB*

Bringing It All Back Home

We Canadians are changing the way we live: the way we work and the way we play. Granted the "paperless office" has failed to materialize, but we are quickly forging ahead into the Post-Information Society. And how much of our electricity is devoted to providing us with information or entertainment (not to mention information-as-entertainment)? Lots.

Consider, if you will, these last two examples of electricity at work in the world of tomorrow...

Press "Enter" Now

In 1990, Bell Canada launched a new videotex system called ALEX, modelled on the extremely popular Minitel phone network that has been operating in France for several years.

It works like this: for a small monthly rental fee, you get a personal computer terminal from Bell to hook into your phone line. With it you can access almost 600 different on-line data services around the country. Many of them are free. The range of available information is enormous, and growing daily: from agricultural and financial news to home shopping classified ads. Nor is the communication just one-way. With ALEX you can find a friend on a chat-line, book a vacation, sign up for a correspondence course, or even start your own videotex service!

Telephone as futurephone: communicating in new ways with ALEX. *Courtesy of Bell Canada.*

Courtesy of Superstock.

Television à la Carte

Telephones and computers are not the only interactive wave of the future. Television is going the same route.

In January of 1990, Télé-Métropole of Quebec premiered the world's first interactive television broadcast, of a hockey match between the Montreal Canadiens and the Quebec Nordiques. With a special converter from Télé-Métropole's sister company, Vidéoway, viewers could push a button for instant replay any time during the game. Other buttons let them choose between cameras following key players.

In March, the station began offering an interactive version of the evening news that offers viewers the possibility of cutting away to in-depth coverage of particular stories, or following the regular newscast.

Can you imagine watching TV in the 21st century?

The Last Word...

We've come a long way in our journey through *The Electric Century.* We've looked at electricity's impact on society, society's impact on electricity, and the world's overall development.

We are surely the heirs to good fortune in Canada's bountiful resources that make the creation of vast quantities of electricity possible. The amount of human toil and effort that has been required to harness this potential and steward it wisely is equally great. We have a proud history of Canadians working together on one breakthrough after another, and accomplishing great feats.

Upon reading this book, it's hard not to think about the role electricity has played in our lives. The homes we live in and all the things we enjoy seem virtually to come to us courtesy of electricity. Our jobs likely require electric power for some critical process on a day-to-day basis.

Our earliest electric beginnings in Canada date back to 1846, some 145 years ago. The Canadian Electrical Association was founded one hundred years ago.

Now a new century is dawning. Its shape is in our hands.

Courtesy of CEA.

THE CANADIAN ELECTRICAL ASSOCIATION 1891 - 1991

Our story would not be complete without due recognition of the history of the Canadian Electrical Association itself, especially in this the centennial of its birth.

It all began, simply enough, with a conversation. In January, 1891, a new magazine called *Canadian Electrical News* was started by a Toronto publisher, C.H. Mortimer. The managers of several electrical companies had been bandying about the idea of forming an industry association, and they came to him with the idea.

Mortimer was interested, and in July he wrote an open letter to the "owners of central stations and other electric men" proposing a meeting to discuss forming just such a group.

The response was rapid and enthusiastic: on September 17, 1891, 28 people showed up for an organizing meeting at the Toronto offices of the Industrial Exhibition Association. Twenty came representing eleven different utilities, four were manufacturers of electrical equipment, three were from telephone companies, and one came on behalf of the Electric Lighting Board of Underwriters.

J.J. Wright, president of the Toronto Electric Light Company, was asked to preside, and C.H. Mortimer acted as secretary.

Needless to say, the meeting was a great success. A seven-member organizing committee was struck, and on the 26th of November, with constitution and bylaws in place, the Canadian Electrical Association was born. Its mandate: "To foster and encourage the science of electricity and promote the interests of those engaged in any electrical enterprise."

J.J. Wright was voted President. K.J. Dunstan, of Bell Telephone, Toronto, and John Carroll, of the Eugene Phillips Electrical Works, Montreal, became first and second vice-presidents, respectively, and C.H. Mortimer was appointed as Secretary-Treasurer. A nine-member executive committee was also established.

The Early Years

The association's first annual convention was held June 14-16, 1892, in the offices of the Hamilton Board of Trade, with sixty people present. It's interesting to note that right from the beginning, the

CEA has represented not only utilities, but also manufacturers of electrical equipment and related industries, as seen in the affiliations of its first officers. Indeed, 23 participants at the first convention worked for telephone companies!

By the end of its first year, the association had 55 full members (including ten from western Canada, nine from Quebec, two from Nova Scotia), as well as 24 associate members.

In 1894, the fourth CEA convention was held, outside Ontario for the first time, in Montreal, and the association boasted 137 members.

At that event, participants lamented the fact that, in Ontario, investment of private capital in electric power plants had been replaced largely by government ownership, and went on record saying, "It is only natural that the chief activities of the Association should have been transferred to Quebec and other provinces where there was and is greater scope for useful work."

J.J. Wright, first president of the CEA, from an 1895 photograph. *Courtesy of CEA.*

Government control continued to be a hot issue at the next five conventions as well, as did competition from gas companies.

By 1911, membership had climbed to 500, and the CEA joined forces with an American group, the National Electric Light Association, in order that Canadian operators might better benefit from the experiences of their southern cousins (although this link was terminated in 1933 when NELA disbanded).

On March 30, 1922, CEA was granted its Letters Patent by Dominion Charter.

Increasing the national base of the association was also a major focus in these years, and to that end the first convention in the Maritimes was held at St.-Andrews-by-the-Sea, New Brunswick, in 1929. The first western convention followed at Banff, Alberta, in 1931.

New Horizons

The CEA has undergone many developments in the course of Canada's electric history. One of the association's most important transformations has involved evolution towards a broader constituency and constant attention to refining its programs to meet the needs of public interest in changing times.

First, in 1940, membership in the association was changed from an individual to corporate basis.

Courtesy of CEA.

In 1958, Provincial Power Commissions became eligible for corporate membership in the association. The same franchise was extended to Municipal Electrical Utilities in 1964. These developments had tremendous impact because they enabled the CEA to reflect Canada's unique combination of public and private utility ownership more accurately.

Membership in CEA now includes 34 corporate utility members, 40 corporate manufacturers and more than 2,400 individuals. Today, CEA represents an industry with assets of $100 billion and employs more than 90,000 Canadians.

Over the years, the CEA has also forged important communication links with many other groups, such as the Electrical and Electronic Manufacturers Association of Canada (EEMAC), the Canadian Standards Association (CSA), the Municipal Electrical Association (MEA), the Institute of Electrical and Electronics Engineers (IEEE), and other electrical associations around the world.

Today, the Canadian Electrical Association is truly the national forum and voice of the electrical industry in Canada.

A Sense of Mission

Here, in their own words, is the mission that the CEA has taken up:

- To be a national association which improves the effectiveness of the Canadian

Left and centre, courtesy of CEA. Right, courtesy of Hydro-Québec.

electrical utility industry and fosters better understanding of the industry and its contributions to the quality of life of Canadians.

- To provide a forum for the study, discussion and exchange of information related to the electric utility industry.
- To develop and promote policies and programs, including research and development, respecting production, distribution, safe and wise use of electricity, and protection of the environment.
- To inform the public, government, and business of the responsibilities, goals, and achievements of the Canadian electric utility industry.
- To monitor current and potential issues of national concern to the industry and act as an industry advocate at the national level.
- To facilitate Canadian involvement in international activities affecting the electric utility industry.

Leadership in a Changing World

From its humble beginnings one hundred years ago, the Canadian Electrical Association has grown to become one of the most respected organizations of its kind in the world.

That leadership has brought with it a wide range of responsibilities, not the least of which is a commitment to openness and education, as well as excellence in innovation.

Today, to meet those goals, the CEA has matured into a complex organization with many diverse activities.

For example, within its engineering and operating division are committees on outage statistics, environmental coordination and engineering education, as well as sections on distribution, health and safety, hydraulic power, metering, right of way, transmission, construction methods and equipment, electrical apparatus, electrical coordination, power systems planning and operation, telecontrol and telecommunications, and thermal and nuclear power.

In the research and development division can be found committees on generation, transmission systems, distribution, and utilization. In fact, R&D has become one of the CEA's biggest undertakings, involving more than one

Experimenting with superconductivity at IREQ. *Courtesy of CEA.*

hundred ongoing projects each year, at an annual cost of more than $10 million.

In its corporate resources division can be found committees on performance and productivity evaluation and electric power statistics, as well as sections on finance, computing and information processing, human resources, law, materials management, risk management, and insurance.

In the customer services division can be found sections on communication and public relations, customer service policy, marketing and energy efficiency, rates and load studies, as well as energy economics and forecasting.

People, Power, Progress

The theme CEA has chosen for this, its centennial, is: "People, Power, Progress."

It is both a fitting tribute to the remarkable achievements of the past, and a promise of continuing commitment to a brighter future for all Canadians.

TIMELINE

1267	Roger Bacon publishes his theories on the polar attraction of magnetite.
1600	Dr. William Gilbert, court physician to HRH Queen Elizabeth I, publishes treatise linking electricity and magnetism.
1745	Musschenbroeck gives a demonstration of the "Leyden Jar," precursor to the condenser.
1746	Watson passes an electric current 10,600 feet down a wire.
1747	Benjamin Franklin announces results of experiments on electricity in nature.
1751	First experiments with telegraphy.
1752	Benjamin Franklin demonstrates with a kite that lightning is electricity.
1800	Volta invents the battery.
1809	Sir Humphrey Davy lights the first arc lamp, using a 2,000-cell battery.
1820	Ampère explains electric telegraphy.
1821	Michael Faraday demonstrates that voltage could be generated by moving a wire through a magnetic field.
1825	Sturgeon builds an electro-magnet.
1831	Faraday builds first dynamo, supplying steady electricity.
	Joseph Henry announces principles of electro-magnetic telegraphy, builds first electric motor.
1834	Morse builds an experimental electro-magnetic telegraph.
1837	Montreal Gas Co. & Quebec Gas Co. get exclusive rights to street lighting in those cities.
1838	Morse obtains patent for electro-magnetic telegraph.
1844	Morse builds first telegraph line (Washington to Baltimore).
1846	Toronto, Niagara & St. Catharines Electro-Magnetic Telegraph Co. founded.
	December 19. First Canadian telegram sent between Toronto and Hamilton.
1847	Montreal Telegraph Company founded, builds lines from Quebec City to Toronto.
1853	April 16. Toronto Locomotive Works completes the *Toronto*, the first steam locomotive built in Canada.
1855	June 23. Laying of first transatlantic cable commenced from Nfld. Failure.
1856	Frederick Newton Gisborne lays first successful North American underwater cable (from Nfld. to Cape Breton).

1858	August 2. First transatlantic cable, from Valentia, Ireland, landed at Bay Bulls Arm, Nfld. First transatlantic cable sent from HRH Queen Victoria to President Buchanan (85 words). Line failed after a few weeks.
1866	July 27. First successful transatlantic cable landed at Heart's Content, Nfld., by the *Great Eastern*, the world's largest ship. First successful telegraph link between old world and new.
1870	Gramme patents an industrial dynamo.
	Ferranti begins work on a high-tension AC system in London.
1873	First arc lamp in Winnipeg installed, in front of the Davis Hotel.
1874	July 26. Alexander Graham Bell discloses for first time his conception of the principle of the telephone.
1876	March 10. Alexander Graham Bell makes first phone call in Boston.
	August 10. Bell makes first long-distance phone call over Dominion Telegraph lines. Brantford to Paris, Ontario, with battery power from Toronto. Total distance: 218 km.
1877	Bell turns over 75 percent of patents to his father and Charles Williams, a Boston manufacturer. Williams can't handle obligation to make 1,000 phones, bows out. James Cowherd, from Brantford, sets up Canada's first phone factory. By 1881 he'd made 2,398 phones.
	June 20. First commercial phone service in Canada begins in Hamilton, Ont.
1878	Improved arc lamp, known as the Jablachkov 'candle', demonstrated at Paris Exposition. Montreal businessman J.A. Craig brought one back from Paris and demonstrated it to the Jesuit Fathers of Montreal in 1879.
	Canada's first phone exchange, in Hamilton. 7 lines, 50 subscribers. Phone exchanges set up in same year in Nfld., N.B., Winnipeg and Victoria.
	American Electric & Illuminating Co. founded in Montreal. First of its kind in Canada.
1879	Thomas Edison invents an incandescent lamp that burns for 40 hours.
	Bell's father tries to sell company. No Canadian takers. Sells to National Bell (U.S.) in 1880.
1880s	Numerous electrical companies in Montreal and Quebec couldn't get street lighting contracts. Serviced hotels, stores, banks.
1881	Halifax Electric Light Co. founded in N.S.
	J.J. White builds Toronto's first generator, for T. Eaton Co.
	Ahearn & Soper put industrial lighting into operation in Ottawa.

1882 September 4, 3 p.m. Edison throws the switch at the world's first commercial generating station, on Pearl Street, and distribution system for parts of Manhattan.

Ottawa River mills using incandescent lights.

Introduction of carbon arc street lamps in Winnipeg, from Manitoba Electric Light & Power Co.

F. Nicholls founds Toronto Electric Supply Co. Later becomes CGE.

August 23. Pile O'Bones is officially renamed Regina when the CPR tracks arrive.

Robert Burns McMicking experiments with electric street lighting in Victoria, B.C.

1883 Parliament Buildings in Ottawa lighted by incandescent lights, powered by steam-driven dynamo in plant on edge of Ottawa River.

Victoria, B.C., contracts with Robert Burns McMicking to build and install three arc lamps of 50,000 cp each, on 50 metre masts.

First streetcar demonstrated at Toronto Industrial Exhibition (later CNE).

Northwest Mounted Police use first phones in Regina, Sask.

Hamilton installs first incandescent street lights in Canada.

1884 Canadian Cottons Ltd., Cornwall, installs lighting system designed by Edison in one weave shed. Wilbur Hitchcock assists, later starts Cornwall Street Railway, Light and Power Co. Ltd.

First operating streetcar at CNE. Problems with voltage drop resulting from DC transmission.

St. John Electric Light Co. founded in N.B. Power capacity of 2,000 16 cp lights.

Telegraph and telephone links throughout most of Canada.

Royal Electric (forerunner of Montreal Light, Heat and Power) created in Montreal.

Pembroke Electric Light Co. established by R.B. McAllister, in Ont. First public electric utility in Canada. Still operating.

First electric street lights installed in Toronto.

1885 St. John's Electric Light Co., Nfld., installs steam generator.

January 24. CPR trans-Canada telegraph line completed.

November 7. Lord Strathcona drives last spike on CPR trans-Canada railway, at Craigellachie, B.C.

1885 Electric sewing machines hit the market.

Sept. 30. Lieutenant-Governor Masson inaugurates new electric light system on Dufferin Terrace in Quebec City. 34 arc lamps. Power from Montmorency Falls.

Halifax inaugurates first electrical system: power to 75 street lights and 50 stores.

Ottawa powers street lights with hydro-electricity.

1886 40 percent of energy from animal power; 40 percent from machines; 20 percent from human power. Now 99 percent from machines.

November 4. Forty-six businesses in St. John's, Nfld., using electricity and first street lighting contract signed with Nfld. government.

Prince Edward Island Electric Co. founded. First power plant in P.E.I. Steam-driven, using coal, with a 150 kW generator.

Montreal bows to public pressure and adopts electric street lights. Three tenders entered. Royal Electric Co. wins in Montreal. Quebec Railway, Light & Power Co. wins in Quebec. Soon merged with their competitors. Royal and Montreal Gas merge again with two other electrical companies to form Montreal Light, Heat & Power Co. in 1901.

September 13. Official opening of CP Telegraph system.

December. Telegraph message sent from New Westminster, B.C., to Canso, N.S. Took 3 minutes.

May 19. Victoria Electric Lighting Co. and Vancouver Electric Light Co. both founded. Forerunners of, in turn, B.C. Electric and then B.C. Hydro.

1887 Elihu Thomson invents induction-repulsion AC motor.

Colonel R.W. Leonard, an engineer from Brantford, Ont., builds a hydro-electric factory on Twelve Mile Creek, near Welland.

June 14. The first CPR steamship from Japan arrives in Vancouver, opening direct contact with the Pacific Rim.

1888 Nikola Tesla invents polyphase AC current system and motors.

Victoria agrees to build street railway system. Service begins in 1890.

1889 First street lights in Calgary, Alta. Powered by lumberyard sawdust. 75 kW steam generator.

Arc lamp street lights installed in Brandon and Winnipeg.

First single-phase AC generators in Canada, in Calgary at Bow River Lumber Company and in Ottawa at Chaudière Electric's generating plant.

1890 Lights on in Saskatchewan: Regina, Moose Jaw, Prince Albert. Moose Jaw system runs 350 16 cp bulbs: $1.50 per light per month rates.

1891 Canadian Electrical Association founded.

Victoria Electric Tramway & Light Co. boasts 25 kilometres of track and 11 cars.

Edmonton Electric Lighting and Power Company delivers its first electricity three days before Christmas, from coal-fired steam plant on the banks of North Saskatchewan River. Destined to become Canada's largest municipally-owned electric utility.

1892 Winnipeg Electric Street Railway Co. formed.

Thomas Ahearn, of later fame for the Ottawa Electric Railway Co., displays a device for cooking food with electricity, at the Ottawa Exhibition.

Canadian General Electric established.

First hydro-electric installation in Montreal built on Lachine canal.

Canadian Niagara Power Co. founded.

Thomas L. "Carbide" Wilson, of Woodstock, Ont., uses an electric oven to make calcium carbide, an essential ingredient for making acetylene (used in electric welding).

The Almalgamated Transit Workers Union formed.

1893 Power from Niagara Falls (Queenston-Chippawa line) with the world's largest generating station (3 x 5,000 hp).

Hydro generation first used in Calgary, Alta. 280 hp water wheel on Bow River within city limits.

1894 Victoria Electric Railway & Lighting Co., and Consolidated Railway Co. formed in Victoria, B.C.

1895 First 3-phase AC plant in Canada. St. Hyacinthe, Que. Power plant at Rapid Plat, 9 km from the city. Using three CGE 150 kW, 2,500 volt, 60 Hz, 3-phase alternator.

Halifax Electric Tramways purchases Nova Scotia Power, Halifax Street Railway and Halifax Electric Illuminating & Motor companies.

Quebec Railway, Light, Heat & Power formed by mergers. Later known as Quebec Power Co.

1896 December 5. First "horseless carriage" appeared in streets of Toronto. Electric car built for F.B. Fetherstonhaugh by Dickson's Coachworks of Toronto. Travelled at a maximum speed of 15 mph.

1896 Streetcars in Halifax, N.S.

Guglielmo Marconi demonstrates radio waves. Transmits 4 miles in England. Obtains patent.

July 21. First public screening of a motion picture in Canada.

1897 Shawinigan Water & Power Corp. formed in Que.

First long-distance, high-tension transmission (11 kV) of electric power in British Empire by North Shore Power Co., from St-Narcisse on Batiscan River to Trois-Rivières, Que.: 27 km.

Lots of small networks in operation in Que.

Westinghouse opens a Canadian branch office in Hamilton.

B.C. Electric Railway Co. takes over Victoria Electric Railway Co. and Consolidated Railway Co.

Canadian National Electrical Code first published.

1898 First hydro-electric plant in Newfoundland built at Black River, Placentia Bay, by the Newfoundland Pulp Company.

B.C. Electric Railway Co. build's B.C.'s first hydro plant, 32 km from Victoria.

West Kootenay Power builds B.C.'s second hydro plant in same year, at Bonnington Falls.

Cataract Power Co. completes De Cew Falls hydraulic plant near St. Catharines, transmitting 56 km to Hamilton, Ont. at 11,000 volts. Oldest high-head station in Canada.

Because of a strike, U.S. puts export embargo on charcoal.

Royal Electric Co. produces 20,000 hp in Chambly, transmits it 26 km to Montreal.

1899 Troops are called out in London, Ont., to quell a riot sparked by a streetcar strike.

1900 St. John's Street Railway Co., Nfld., installs first hydro plant for public use in Nfld., at Petty Harbour, 13 km from St. John's. 1,200 kW power at 15,000 volts.

December 23. Reginald Aubrey Fessenden sends first radio voice message.

First electric arc steel-making furnace introduced.

Installed hydro-electric power in Canada reaches 173,000 hp.

1901 December 12, 12:30 p.m. Marconi flies kite with aerial from Signal Hill, Nfld. and receives first transatlantic wireless message, from Poldhu, Cornwall, in England.

St. Lawrence Power Co. installs arc lamps on Cornwall Canal.

Montreal Light, Heat & Power Co. formed.

1902 United Towns Electrical Company formed to serve towns of Carbonear, Harbour Grace and Heart's Content, in Nfld.

Beauharnois Light, Heat and Power Co. created in Beauharnois, 40 km from Montreal on St. Lawrence River. Destined to become Quebec's fourth-largest hydro generating station.

Fredericton, Newcastle and Loggieville, N.B. power up.

Six separate electrical companies operating in Manitoba.

Winnipeg Electric Street Railway Co. merges with Winnipeg General Power Co. to become Winnipeg Electric Railway Co.

December 5. Marconi sends wireless signal across Atlantic from Glace Bay, N.S.

December 15. Canada's first official transatlantic wireless message sent.

Strike in Pennsylvania coal fields stops shipment of coal to Ont. Major incentive for formation of Ontario Hydro.

For $13,500, the City of Edmonton purchases the Edmonton Electric Lighting and Power Company which becomes a public utility. The power plant is moved to higher elevation to avoid floods that frequently interrupted power.

1903 Shawinigan Electric Co. transmits power at 50 kV to Montreal (136 km) along the world's longest transmission line.

B.C. Electric completes hydro plant at Coquitlam Lake, 24 km from Vancouver.

1904 First hydro plant in N.B., on Meduxnekeag River near Woodstock.

1906 Ontario Hydro-Electric Power Commission established. Adam Beck, former mayor of London, Ont., named Chairman.

Winnipeg receives hydro power from Winnipeg River, 132 km away.

Police called out in Winnipeg Transit Strike.

Fessenden makes first transatlantic voice broadcast in November. Fessenden never got any support from the government. Marconi already had exclusive rights to build radio stations, and $80,000 in assistance. In 1928, Fessenden received $1 million settlement (at age 62). He died in Bermuda at 66, in 1932.

May 28. The Ouimetoscope, Canada's first motion picture theatre, opens in a converted Montreal dance hall.

1907 November 28. Canada's first dial telephones used, in Sydney Mines, N.S.

1909	Calgary Power formed after various mergers & transactions. Develops into largest investor-owned utility in Canada (TransAlta). Now provides two thirds of Alberta's electricity.
	February 23. *The Silver Dart*, flown by J. McCurdy, makes first flight in British Empire, at Baddeck, N.S.
1910	Installed hydro-electric power in Canada reaches 977,000 hp.
	First transmission line (110 kV) from Niagara Falls, goes to Dundas, Ont.
1911	Pointe-du-Bois hydro plant opened, on 165-km circuit to Winnipeg, Man.
	Alberta's second hydro plant built at Horseshoe Falls, on Bow River, 88 km from Calgary.
1912	Steel Company of Canada inaugurates world's first completely electrically operated steel mill.
1913	Alberta's third hydro plant established at Kananaskis, on Bow River.
	First electric reversing mine hoist in Canada.
1915	CNR rail line completed from Que. to Vancouver.
	Lee de Forest, an American, invents the first AC vacuum tube, the "Audion."
1917	Maritime Electric Co. founded. Subsequently grew and acquired numerous local utilities.
1919	Nova Scotia Power Commission established.
	Manitoba Power Commission established.
	November 4. World's first radio station, CFCF, begins broadcasting in Montreal.
	Ernest Rutherford achieves the first induced atomic reaction.
1920	St. John's Electric Light & Power Company formed when street railway and electrical distribution business separated from other interests of the Reid Newfoundland Company.
	New Brunswick Power Commission established.
	Installed hydro-electric power in Canada reaches 2,515,559 hp.
	October 7-17. First trans-Canada flight.
1921	CEA creates Safety Medal program.
	Ontario Hydro opens Sir Adam Beck No. 1, largest power plant in the world at that time.
1922	CEA receives Dominion Charter.
1923	CEA strikes first Safety Medal.

1923 34 radio stations operating in Canada.

March 14. First Canadian radio broadcast of a professional hockey game (CKCK in Regina).

March 22. Foster Hewitt, long acknowledged as the dean of hockey, makes his first radio broadcast, from the Mutual Street Arena in Toronto.

1924 St. John's Light & Power Co. purchased by Canadian industrialist, Izaak Walton Killam, who incorporated Newfoundland Light & Power Co. Ltd. and made it part of International Power.

Winnipeg Electric Railway Co. renamed Winnipeg Electric Co.

October 26. First successful radio-transmission of newspaper photo. Appeared in *London Daily Mail*. System invented by Canadian, Sir William Stephenson.

1925 Ted Rogers invents first radio receiver that could be plugged into an electrical outlet, thanks to improved AC vacuum tube he invents (superior to previously-invented U.S. tube).

1926 October 20. Yukon and N.W.T. linked with south by radio. First successful link. Telegraph had failed due to rugged terrain and conditions.

1927 Ted Rogers' improved batteryless system reaches commercial quality. He starts Radio 9RB (later CFRB) in Toronto.

November 10. Invention of electric organ by Morse Robb announced in *Toronto Daily Star.*

Canadian Electrical Code adopted by Canadian Standards Association.

Canadian Utilities formed in Alta. The company purchases small generating plants and distribution systems, begins serving a number of Alta. communities.

1928 Nova Scotia Light & Power Co. established.

First transmission at 220 kV from Paugan Falls in Que., to Toronto (more than 440 km).

Edmonton Power installs one of the world's first 10 MW turbo-generators designed to run at 3,600 rpm at its Rossdale Plant.

1929 119 generating stations in Saskatchewan, serving 20 percent of population. No rural.

Saskatchewan Power Commission established.

August 26. First diesel electric locomotive in North America completes trip from Montreal to Toronto. Beginning of the end for steam locomotives.

1930 Installed hydro-electric power in Canada reaches 6,125,012 hp.

1932 January 25. First trans-Canada phone system formally inaugurated by the Earl of Bessborough, the Governor General.

1932 May 24. Parliament passes bill establishing a nationally-owned broadcasting system.

1936 November 2. World's first TV program (*Picture Page*, on BBC). First person on screen was Joan Miller of Nelson, B.C.

1937 Nova Scotia passes Rural Electrification Act.

1939 January 10. Inauguration of Canada/Nfld. link via radio. Avalon Telephone Co. & Montreal Bell. This single-channel, AM radio circuit was Nfld's only phone link with Canada and the rest of the world until it joined Confederation in 1949.

Professor Eli Franklin Burton builds North America's first electron microscope in Toronto.

1940 CEA boasts 793 members.

Canadian Utilities Ltd. introduces the mobile radio transmitter-receiver.

1941 Edmonton Power's Rossdale Generation Station, at 330 MW, becomes Canada's largest thermal-power plant with six, coal-fired steam turbine sets. The plant is converted to natural gas in the 1950s.

1944 Quebec Hydro-Electric Commission (Hydro-Québec) formed an expropriation of existing investor-owned utilities. Now one of the ten largest public utilities in North America.

1947 Discovery of oil in Alta. (Leduc No. 1).

1948 93 percent of Nfld's electricity used in pulp and paper industry.

November 20. First microwave phone call in Canada, between P.E.I. and N.S. They were tired of the problems they were having with their undersea cable.

1949 Saskatchewan Power Commission becomes Saskatchewan Power Corporation. Government-owned.

Saskatchewan's Rural Electrification Act passed. By 1960, power reaches 67,000 farms and all villages and hamlets in populated half of province.

1950s Natural gas discovered in Sask.

Arrival of western gas in eastern Canada.

1950 Calgary Power builds prototype underground cable distribution system for subdivision on Crestview Road. Key factor is using polyethylene insulation rather than rubber or PVC. Instant national and international success.

Installed hydro-electric power in Canada reaches 12,562,750 hp.

1951 Dr. John Hopps in Ont. completes work (with National Research Council) on world's first pacemaker. Prototype uses a car battery.

1951 First 100 MW steam turbo-generator set in Canada placed in-service at the Richard L. Hearn plant in Toronto.

1952 September 6. First television program in Canada broadcast in Montreal.

Atomic Energy of Canada Limited formed.

October 11. First television hockey broadcast, in French, on CBFT in Montreal.

November 1. First television hockey broadcast, in English, on CBLT in Toronto.

1953 Preliminary studies begun for NPD, Canada's first nuclear power plant. The project was approved in 1955. Construction began the next year. The reactor went critical in 1962, and first electricity was fed into the Ontario Hydro grid on June 4, 1962.

1954 Newfoundland & Labrador Power Commission formed.

Alcan (Kemano-Kitimat) transmits electricity at 287 kV.

1955 October 5. Canada announces plans to build nuclear research reactor for India under Columbo plan.

1957 B.C. Electric first utility in Canada to build 360 kV transmission line.

First central pumped water storage for electric power generation in Canada put into service at Sir Adam Beck No. 2 plant, Niagara Falls. Each of the six units capable of producing 29,450 kW when reversed.

1958 Completion of world's longest microwave link, from Halifax to Vancouver. 8,500 km; 139 towers. Eventually carrying 1,200 channels.

1959 Saskatchewan Power Corporation begins construction of first major hydro plant at Squaw Rapids.

Ontario Hydro builds first test line for 3-phase, extra-high tension (500 kV).

Hydro-Québec completes one of its first megaprojects: Bersimis I Hydro Generating Station on Bersimis River in Manicouagan region, after seven years of work.

1960 Hydro-Québec completes Bersimis II Hydro Generating Station.

Installed hydro-electric power in Canada reaches 26,688,094 hp.

There are more than 80 different kinds of household appliances on the market.

July 21. Announcement of Canada's second nuclear research centre: Whiteshell, on Winnipeg River, 130 kilometres northeast of Winnipeg.

1961 Hydro-Québec completes redevelopment at Beauharnois Generating Station, making it Quebec's fourth-largest installation after the giants of James Bay.

First generation at 300 MW in Canada, at Lakeview station.

1962 Calgary Power puts two 15 kV, fully-articulated manlifts into operation. Allows for direct hand-contact and "hot-line" work with lines over 5 kV. Greatly speeds up line work; equipment pays for itself in 1-2 years. First serious trial in Canada. Instant success.

SaskPower installs first underground distribution system for housing subdivisions, at 25 kV.

September 28. Canada launches Alouette I, scientific satellite. Third country in the world to launch a satellite.

Canada's prototype nuclear generating station begins operation at 20 MW.

Pembroke Electric Light Co. inaugurates Canada's first radio remote control system for unattended generating stations, as well as first use of prestressed concrete pilon footings.

1963 Hydro-Québec acquires most of the province's remaining private and cooperative electrical installations.

1964 CNCP completes their own microwave system.

Hydro-Québec begins building 735 kV transmission line from Manic II, III and V at Manicouagan-Outardes project to Montreal (600 km away) with world-class technology.

Hydro-Québec completes Carillon Hydro-Electric Generating Station on the Ottawa River.

August 20. Plans for Pickering Nuclear Generating Station announced.

1965 Canada launches Alouette II, scientific satellite.

Newfoundland Power Commission renamed Newfoundland & Labrador Power Commission.

November 15. Douglas Point Nuclear Generating Station (200 MW generating capacity) goes critical. First electricity to Ontario Hydro's grid on January 7, 1967.

1966 Newfoundland Light & Power Co. Ltd. formed with the amalgamation of five privately-owned electric utilities.

1967 Saskatchewan Power Corporation builds second major hydro plant at Coteau Creek.

1969 Canada launches Isis I, scientific satellite.

Telecom Canada formed.

1970 Edmonton Power's Clover Bar Generating Station is the first facility in North America with spring-supported foundation table under turbine generators. Fourth 165 MW gas-fired unit completed in 1978.

1971 Canada launches Isis II, scientific satellite.

First unit of Pickering Nuclear Generating Station goes into service.

First two units of Churchill Falls Hydro-Electric Generating Station achieve full power. When all eleven units (of 475,000 kW each) were completed in 1974, Churchill Falls became the largest hydro-electric installation of its kind in the world. This honour went to La Grande II in Quebec's James Bay project in 1978.

Nova Scotia Light & Power Co. merges with Nova Scotia Power Commission to form Nova Scotia Power Corp.

November 9. Anik A-1 communications satellite launched. Begins broadcasting January 11, 1973: 12 TV channels, 11,520 voice circuits. Reaching 25 northern communities by spring.

1972 Alberta Power comes into being as the operating company for all Canadian Utilities' electric holdings, including Yukon Electric.

1973 April 20. Anik A-2 communications satellite launched.

1975 May 7. Anik A-3 communications satellite launched.

Newfoundland & Labrador Power Commission renamed Newfoundland & Labrador Hydro.

November 11. James Bay Convention signed by Hydro-Québec and area Cree.

1977 First test of fibre optics for phone trunklines between two stations, in Montreal. Northern Telecom and Bell–Northern Research (a subsidiary) now world-leader in this technology.

1978 Anik B communications satellite launched.

1981 Manitoba Tel tests world's first rural fibre optics transmission system. Project Ida, named after Manitoba's first phone operator.

Hydro-Québec's La Grande II, Quebec's largest hydro station (with more than twice the installed capacity of its runner-up, La Grande IV), commissioned. Largest underground generating station in the world.

1982 Anik C-1 launched by space shuttle *Columbia*.

Anik D-1 launched. Made by SPAR Aerospace. Doubles communication capacity of A series.

N.B. commissions 630,000 kW CANDU reactor at Point Lepreau. Reaches 50 percent power by year end.

1983	Point Lepreau nuclear station reaches full power; accepted into commercial operation.
	Bell Canada using fibre optics exclusively for all new inter-office trunk cable requirements.
	SaskPower begins rebuilding rural 14.4 kV system underground.
1984	August 25. Nova Scotia Tidal Power Corp. commissions first tidal generating station in North America, using 20 MW Straflo Turbine with rim-mounted generator.
	SaskTel completes world's longest fibre optics system (3,400 km). A fibre optics pair carries 4,000 messages simultaneously, compared with 2 on a copper pair.
	Direct-broadcast satellite TV beamed into home dishes.
	Hydro-Québec's La Grande III Generating Station (Quebec's third largest) completed.
1985	Telecom Canada promises coast-to-coast fibre optic network by 1995.
	Anik C-2 and C-3 communications satellites launched.
	Newfoundland Light & Power celebrates centennial, as it was on October 17, 1885 that electricity for public use was first generated in Nfld. by the St. John's Electric Co., one of its predecessor companies.
1986	Hydro-Québec's La Grande IV Hydro Station (Quebec's second largest) completed.
	Hydro-Québec builds Kuujjuaq Wind Turbine Generating Station demonstration project.
1988	Hydro-Québec serving more than 3 million customers.
1989	Alberta Power commissions a high-voltage DC tie-line linking the electric grids in Alta. and Sask. This represents an electrical "Last Spike" – previously, the electric systems in the eastern and western halves of Canada had not been connected.
	B.C. Hydro introduces new PowerSmart conservation program.
	Edmonton Power delivers first 400 MW unit of coal-fired Genesee Generating Station to Alberta Interconnected System. Edmonton Power had begun to purchase coal leases in the area, 50 km south-west of the city, in the late 1950s.
1990	"There's a future in Science and Technology/Un bel avenir en perspective. Les sciences et la technologie." campaign launched by Industry, Science and Technology Canada.
	Bell Canada inaugurates ALEX videotex service.
	Vidéoway Ltée inaugurates interactive television service.
	Newfoundland Light & Power Co. Ltd. now known as Newfoundland Power.
1991	Centennial of the formation of the Canadian Electrical Association. The future begins.

SUGGESTED READING

B.C. Hydro Story, The, B.C. Hydro (unpublished manuscript), 1973

Bowers, Brian, *A History of Electric Light and Power*, London: Peter Peregrinnis (in association with The Science Museum), 1982

Canby, Edward Tatnall, *A History of Electricity*, New York: Hawshorne Books, 1963

Clark, Ronald W., *Edison: The Man Who Made the Future*, New York: G.P. Putnam's Sons, 1977

Conot, Robert, *A Streak of Luck: The Life & Legend of Thomas Alva Edison*, New York: Seaview Books, 1979

Donaldson, Gordon, *Sausages, Schnitzels & Public Power: A Brief History of Ontario Hydro's First 75 Years*, Toronto: Ontario Hydro, 1981

Energy and Canadians: Into the 21st Century, Energy, Mines and Resources Canada, 1988

Hawkins, W.E., *Electrifying Calgary: A Century of Public & Private Power*, Calgary: University of Calgary, 1987

King, W. James, *The Development of Electrical Technology in the 19th Century*, Washington: United States National Museum, Bulletin 228, Paper 30, 1962

Maclaren, Malcolm, *The Rise of the Electrical Industry During the Nineteenth Century*, Princeton: Princeton University Press, 1943

McKay, Paul, *Electric Empire: The Inside Story of Ontario Hydro*, Toronto: Between The Lines, 1983

Pelletier, Michel, ed., *River Tamers: Québec's Expertise in Hydroelectricity*, Montreal: FORCES, 1989

Prevey, W. Harry, ed., *Electricity: The Magic Medium*, Thornhill: Institute of Electrical and Electronics Engineers Canadian Region, 1985

Sexton, Jack, *MONENCO: The First 75 Years*, Montreal: Monenco Ltd., 1982

Sharlin, Harold I., *The Making of the Electrical Age: From the Telegraph to Automation*, New York: Abeiard-Schuman, 1963

Wyatt, Alan, *Electric Power: Challenges and Choices*, Toronto: The Book Press, 1986

Wyatt, Alan, *CEA Centennial Book*, unpublished manuscript, 1989

White, Clinton, *Power for a Province: A History of Saskatchewan Power*, Regina: University of Regina, 1976